The Origins of Language

Joanna Dornbierer-Stuart

The Origins of Language

An Introduction to Evolutionary Linguistics

palgrave
macmillan

Joanna Dornbierer-Stuart
School of English
Birmingham City University
Birmingham, UK

ISBN 978-3-031-54937-3   ISBN 978-3-031-54938-0 (eBook)
https://doi.org/10.1007/978-3-031-54938-0

Cover credit: Front cover drawing by William Duffy

This Palgrave Macmillan imprint is published by the registered company Springer Nature Switzerland AG
The registered company address is: Gewerbestrasse 11, 6330 Cham, Switzerland

If disposing of this product, please recycle the paper.

To my twin brother, Andrew, my first accomplice in mischief with whom I embarked on the discovery of language. To this day, you inspire me with your boundless zest for life.

Preface

Why study the origins of language? Probably for the same reason that we would investigate anything we do not have complete knowledge about. It is human nature to do so. Language is itself a subject worthy of study. Not only is it undoubtedly the most sophisticated form of communication in the animal kingdom, but it defines who we are. Learning about language equates to learning about humans, and learning about how language developed equates to learning about how the human species developed.

We are first and foremost characterised by our large and highly developed brains, which have enabled us to develop advanced tools, culture and language. Our inquiring minds and curiosity about the environment have led to great technological advancements, and our language provides us with a unique form of communication that allows us to transmit our knowledge to future generations. While all animals communicate, the combination of our special brain design and collaborative behaviour has allowed us to excel in the field of communication. However, the question of how language came about is incredibly complex and can be approached from many angles. It is the aim of this book to offer a broad introduction to the rapidly expanding field of evolutionary linguistics, not only to students trained in languages and linguistics but also to those from other disciplines such as biology, psychology, cognitive science and artificial intelligence who wish to learn more about the complexity and development of human language (as well as human beings).

To understand the origins of language, we first need to consider what language is. There are different views on this. At one end of the spectrum, we have the Darwinian evolutionists, who see human language primarily as an innate biological faculty passed on to every one of us in our genes. Steven Pinker, Professor of Psychology at Harvard University, writes in his widely acclaimed bestseller *The Language Instinct*:

> "There is no reason to doubt that the principal explanation for human language is the same as for any other complex instinct or organ ... A language instinct unique to modern humans poses no more a paradox than a trunk unique to modern elephants." (Pinker 1994, 333, 342)

An underlying assumption of such an innate faculty is that of natural selection, with incremental changes taking hold through fitness advantages. As far as language is concerned, each successive stage in its evolution would have meant humans were better equipped to do such things as negotiate the environment, form alliances and convince others when it came to making decisions that were crucial for survival.

At the opposite end of the spectrum, we have the culturalists, who see very little about human language as innate or shaped by natural selection. Rather, language is culturally constructed and evolves as it is transmitted across communities. According to the pioneering developmental psychologist Michael Tomasello in his influential book *Constructing a Language: A Usage-Based Theory of Language Acquisition* (2003), children learn the language of their community through a mix of imitation and higher-level abstract reasoning and problem-solving strategies. This leads to "the gradual discovery and construction of linguistic knowledge". Following this theory, language structures emerge simply through using language, and children build their language based on general cognitive skills and learning mechanisms, without resorting to some mysterious "language instinct".

It is a main concern of this book to clarify such opposing positions and unify perspectives. Language is unique precisely because it is both a biological capacity and a cultural phenomenon, but the relationship between the two is not a direct one and therefore open to much debate. Nevertheless, what is clear to all evolutionary linguists is that, as in any serious scientific enterprise, theories need to be backed up with observable, objective data. The quest for a comprehensive model embracing all aspects of the origins of human language continues, with some interesting new inputs outlined in this book.

It is my hope that this book will open up to you the fascinating multidisciplinary subject of evolutionary linguistics and make you want to explore more. I hope you will find the book stimulating, eye-opening and enjoyable.

St. Gallen, Switzerland   Joanna Dornbierer-Stuart

Acknowledgements

With many thanks to the following for permission to use copyrighted images:

CartoonStock Ltd. 2023 for use of the cartoon "The Emergence of Language" by Leo Cullum;

paleo artist Mauricio Anton for his personal permission to use a copy of his artwork "Australopithecus family".

I am also very grateful to the following for allowing me to adapt figures and/or use extracts from their work:

José-Luis Mendívil-Giró, Professor of General Linguistics at the University of Zaragoza: figure of the synthesis of the biological and cultural evolution of language in *Did language evolve through language change? On language change, language evolution and grammaticalization theory* (Mendívil-Giró 2019);

Simon Kirby, Professor of Language Evolution at the University of Edinburgh: figure of three adaptive systems in language in *Innateness and culture in the evolution of language* (Kirby et al. 2007);

Dr. Oren Poliva: excerpts from *From Where to What: A Neuroanatomically Based Evolutionary Model of the Emergence of Speech in Humans* (Poliva 2015);

Cedric Boeckx, ICREA Research Professor at the University of Barcelona: excerpts from *Reflections on language evolution: From minimalism to pluralism* (Boeckx 2021);

Prof. Angela Friederici, Director and Scientific Member of the Max Planck Institute of Cognitive Neuroscience, Leipzig: excerpts from *Towards a neural basis of auditory sentence processing* (Friederici 2002);

Andrew Radford, Professor Emeritus and Head of Department of Language and Linguistics at the University of Essex: example utterances of one- to two-year-olds in *Syntactic theory and the acquisition of English syntax* (Radford 1990);

Dr. Seán Roberts, lecturer in the Centre for Language and Communication Research at Cardiff University: excerpts from *CHIELD: the causal hypotheses in evolutionary database* (Roberts et al. 2020).

I would further like to acknowledge my charismatic first Spanish teacher, Mr. Glyn Atherton, for opening up the world of linguistics to me at secondary school, as well as Bernard McGuirk, Emeritus Professor of Romance Literatures at the University of Nottingham, for enabling me to set a precedent at Nottingham and study three subjects, French, Spanish and Linguistics; also thanks to my tutors at Birmingham City University for their valuable insights on topics covered in this book, especially Prof. Andrew Kehoe, regarding the design and uniqueness of human language, Dr. Robert Lawson, regarding language variation and change, and Dr. Tatiana Grieshofer, regarding language acquisition; also many thanks to Miyu Yoshimura in Tokyo, who corrected my Japanese.

Finally, I would like to give a special mention to my nephew, William Duffy, who gave up his free time in between his studies at medical school to produce four anatomical diagrams, in themselves works of art and one of which adorns the front cover of this book; also enormous thanks to my friends and family for supporting me in my endeavour and joining me in my reflections on language for many months, and also to my publisher for constant support and assistance during the publishing process.

About This Book

There is a logical progression throughout the book, with later chapters building on knowledge acquired in earlier chapters. Chapter 1 introduces some key concepts and provides a brief overview of the historical development of the discipline of evolutionary linguistics. This is followed by an introduction to the formalisation of language and an outline of the various sources of evidence required to explain the origin of language. Developing an explanatory theory of language evolution is not possible if we lack a description of the object we wish to explain. Chapter 2, therefore, looks at the structure and organisation of language itself, and Chapter 3, the purpose language serves. In Chapters 4 and 5, we branch out from theoretical linguistics and look at the environmental conditions that affected human evolution, and the resulting social and cultural conditions that favoured the development of language. By this stage in the book, it should be clear that biology and culture are interrelated and co-dependent.

Chapter 6 investigates how languages vary in the short term (culturally) and attempts to fit a theory of historical language change into the larger picture of the biological evolution of language. Chapters 7 and 8 turn to psycholinguistics and neurolinguistics to investigate the mechanisms behind using language. The cognitive and neural processes involved in understanding, producing and acquiring speech are explored to see if they can provide any clues as to how language evolved. Chapter 9 asks what human language has in common with other animal language systems and provides an insight into how the comparative approach can be used to determine the pre-linguistic roots of language, using anatomical, neurological and genetic evidence from various animal species. We will see that the comparative approach also serves to highlight what is unique about human language. A concluding chapter aims to summarise approaches and unite all aspects in a unified understanding of how language came to be the way it is.

If you are a student of linguistics, you will probably already be familiar with much of the terminology in this book. If you are approaching language evolution studies from outside of linguistics, you will find that new terms are explained as they arise. If the same terms are repeated later in the book and you need a recap, you will be able to find a further explanation in the extensive glossary at the end of the book.

Contents

List of Figures

List of Tables

Chapter 1
Introduction

The origin of language has for centuries been the subject of fascination, and sometimes of ridicule. Countless hypotheses have been put forward about how, why, when and where language may have evolved. These range from the theory of "divine origin", to the more earthbound notion that language was a human invention, right up to the curious "ta-ta" theory, which claims words arose as the result of tongue movements that mimicked manual gestures—not so ridiculous when you watch the mouth a young child using scissors. In 1866, the Linguistic Society of Paris famously prohibited any further discussion on the origins of language, judging it to be an irresolvable issue. Five years later, Charles Darwin threw a spanner in the works when he applied his general theory of evolution to human beings. His method of investigation, based on empirical observation across species, is a core feature of today's rigorously scientific discipline of **evolutionary linguistics**.

> **Evolutionary linguistics**: Discipline that tries to explain the emergence and subsequent development of language in humans. It deals with the biological evolution of the capacity for language as well as the cultural development of language. Being a scientific discipline that relies on empirical data, its main challenge is that there are no direct physical traces of early human language.

In this chapter, we will proceed with a brief overview of the historical development of the discipline of evolutionary linguistics, followed by a short introduction to the formalisation of language and an outline of the various sources of evidence required to explain the origin of language.

© The Author(s), under exclusive license to Springer Nature Switzerland AG 2024   1
J. Dornbierer-Stuart, *The Origins of Language*,
https://doi.org/10.1007/978-3-031-54938-0_1

1.1   A Brief History of Evolutionary Linguistics

The discipline of evolutionary linguistics has travelled a tortuous path since its beginnings in the nineteenth century. In 1871, when Charles Darwin applied his general theory of evolution to human beings (in his second book *The Descent of Man)*, he suggested that language itself may have evolved through **natural selection**, more specifically through **sexual selection**. He imagined that language arose from something akin to birdsong, to attract the opposite sex, and that competition eventually led to more elaborate vocal behaviour. At the time, however, there was simply not enough evidence to support this astonishing claim.

> **Natural selection**: Key mechanism of evolution whereby organisms that are better adapted to their environment tend to produce more offspring and transmit more of their genetic characteristics to succeeding generations than those that are less well adapted.

> **Sexual selection**: A mode of natural selection in which members of one sex (usually males) compete for access to members of the opposite sex.

In the early twentieth century, Darwinism was given a new lease of life with a reformulation of evolutionary theory. The so-called "modern synthesis" combined classical Darwinian selection theory with Gregor Mendel's ideas on genetic mutation and population-oriented heredity. This huge intellectual achievement coincided with the philosophical wave of positivism, which sought to reduce science to a framework of hypotheses verified purely by observable evidence. In biology, this naturally led to a focus on outward behaviour, which can be directly perceived and measured.

By the 1950s, **behaviourist** theories, which explained behaviour as a response to stimuli in the environment, had become popular in psychology and also in the study of language. In his influential book *Verbal Behavior* (1957), the American psychologist B. F. Skinner claimed that language was a set of habits that could be acquired by **imitation** alone (i.e. it was acquired from the environment, or "nurture"). Some believed that the behaviourist account was far from adequate to explain a phenomenon as complicated as language learning and argued instead for the innate knowledge of language (i.e. it was inherited through the genes, or "nature").

> **Behaviourism**: View that all behaviours, including language, are a set of habits acquired by imitation. The behaviourist approach ignores the mental processes that underlie behaviour and focuses solely on output.

Imitation: Process whereby an individual observes and replicates another's behaviour, allowing the transfer of behaviour without the need for genetic inheritance.

This new biological perspective was promoted in particular by the linguist and neuroscientist Eric Lenneberg, who in his ground-breaking 1967 study *Biological Foundations of Language* argued that language is biologically programmed, especially since there is a "**critical period**" in which it has to be acquired. With the recognition that language learning follows a set schedule, the focus returned to the biological basis of language and the subject of **biolinguistics** was born.

Critical period: In developmental biology, a maturational period in the life of an organism when the nervous system is especially sensitive to certain environmental stimuli in order to develop a skill that is indispensable for the organism's survival. Language acquisition is a period during which language must be acquired in order to insure the development of native competence.

Biolinguistics: Discipline that seeks to provide a biological framework to understand the fundamentals and evolution of language.

The emerging **nativists**, led by the influential linguist Noam Chomsky, argued that grammar itself lies in the **genes**. It was assumed we are born with a certain knowledge of how language works for a number of reasons. Firstly, it seemed unlikely that infants could "pick up" the highly complex rules of language simply from listening to the speech around them. Ninety-nine per cent of the time, parents do not correct the grammar of their children, let alone explain it to them. It would be a thankless task for an adult to explain to a toddler why the following first three phrases are permitted in English but not the fourth:

> The butter melted.
> I melted the butter.
> The butter sizzled.
> *I sizzled the butter.

Yet, all children who speak English "work out" for themselves which verbs the causative rule applies to and which it does not. Secondly, children create sentences they have never heard before; utterances such as *Mummy goed* or *it breaked* show clearly that children are not copying language around them but following rules. Thus, language learning is not merely based on imitation but also on analogy. Our language must embody structural rules that humans use to govern their language production. Thirdly, children make mistakes of their own which are not copied from the environment. All children make similar mistakes which are not usually found in adult speech. As convinced as we may be by the idea that grammar lies in the genes, it is nevertheless very difficult to produce empirical evidence to support such a claim. It is similarly difficult to discredit the truth of the theory of nativism.

Nativism: View that we are born with a certain knowledge of how language works.

Gene: One of many stretches of DNA in a chromosome that contributes to the specification of some trait of an organism.

Biolinguistics has since split into two distinct schools of interest: those within Chomsky's framework of **generative grammar** who are seriously engaged in discovering the basic design of language, believing it to have emerged through sudden and significant genetic change (the saltationists), and those who have the broader aim of linking the emergence of linguistic structure to biology and natural selection, assuming a gradual process, with each stage building upon the previous stage (the gradualists or adaptationists).

Generative grammar: Noam Chomsky's proposed system of syntactic rules that generate all the possible grammatical sentences of a language and from which phonological and semantic structures are subsequently derived.

The **saltationists** argue that humans evolved language following a single **genetic mutation**, or small set of mutations, that occurred after our evolutionary split from other ape species, and so any language-like skills other primates share with humans (such as gestures, vocalisations and rudimentary symbolic communication) are irrelevant to an explanation of human language. In the Chomskyan view, only humans have language because only humans have the specific genetic mutation for language. Other scientists disagree with this position, believing that human language is not the product of a single genetic endowment, but evolved instead through a series of small, incremental changes over an extended period.

Saltationism: View that language emerged in a single and, in evolutionary terms, sudden step.

> **Genetic mutation**: Change in the DNA sequence of an organism. Means by which nature creates variation in species.

The **adaptationists**, sometimes called neo-Darwinists, view language as a complex **adaptation** that must be studied from the perspective of evolutionary biology. They believe that the special properties of language evolved in stages spanning millions of years. The process would have started with a simple system of "**protolanguage**" that progressively developed into ever-complex systems until language as we know it emerged. American linguist Ray Jackendoff has suggested that in an early stage, individuals would have been able to use sounds symbolically to name a wide range of objects and actions in the environment. A major development would have been the ability to combine sounds in novel ways to create a larger vocabulary. The next step would have been to juxtapose words to create a message. A final series of changes would have added grammatical devices (such as tenses and linking words) to enrich the structure and meaning of messages. Some hypothesise that this last phase could have been a purely cultural development, through imitation and teaching, while others think this, too, required genetic changes in the brains of speakers [1].

> **Adaptationism**: View that human language evolved slowly through genetic adaptation and natural selection.

> **Adaptation**: Trait in a species that has evolved over generations in response to a new environment.

> **Protolanguage**: A hypothesised intermediate stage in the emergence of language that was present in archaic humans.

In the twenty-first century, there has been renewed interest in language as a cultural phenomenon, with the result that the nature-nurture conflict has been reopened under the guise of genes versus culture. In this context, culture refers to immaterial culture, the set of behaviours and knowledge acquired socially, through imitation and learning. The **culturalist** stance highlights the fact that language is not a "crystallised" biological structure but fluid and ever-changing. In *The Descent of Man* (1871), Darwin already noted that language has an evolutionary dynamic of its own that proceeds independently of human biological evolution:

"A struggle for life is constantly going on amongst the words and grammatical forms in each language. The better, the shorter, the easier forms are constantly gaining the upper hand, and they owe their success to their own inherent virtue." [2]

In his book *The Unfolding of Language* (2005), the mathematician turned linguist Guy Deutscher shows how an elaborate system of grammatical rules could have evolved from very humble beginnings through progressive improvements as it is passed on, as a response to the needs of efficient communication [3]. However, just as it is difficult to prove that grammar lies in the genes, so it is equally problematic to simulate the path of a language from scratch. The debate continues (sometimes to the point of becoming emotionally charged) between those who believe grammar is coded in our DNA and those who believe it can be more plausibly explained as the product of **cultural evolution.**

Culturalism: View that language is culturally constructed and socially transmitted. In its strongest form, it can account for the genesis of language.

Cultural evolution: Language change through social transmission rather than genetic inheritance.

According to neurocognitive scientist Dr. Jonas Nölle at the University of Glasgow, neither the cultural evolution of language nor the theory of a biological basis for language can be refuted. Most researchers would certainly acknowledge that some biological structure is required in the first place to start a cultural evolution process. On the other hand, there is growing evidence that linguistic structure can emerge without any changes in biological "hardware". Nölle argues that we need to recognise that **language evolution** is a co-evolved continuum, involving biological, cultural and social forces [4].

Language evolution: Emergence and subsequent development of language in humans. A co-evolved continuum involving biological, cultural and social forces.

Above all, evolutionary linguistics needs to be compatible with the scientific method. Computational linguist Willem Zuidema at the University of Amsterdam has long stressed that in order to achieve a complete and scientifically rigorous model of language evolution, theories need to be formalised so that they can be empirically tested for both internal coherence and consistency with competing theories. To do this, he claims we should be guided by the methodology of evolutionary biology, which draws on observable evidence across species and also makes extensive use of mathematical modelling [5]. Computers are now routinely used in the field of language evolution to quantify the huge amount of data amassed, and mathematical models are employed to develop and test precise hypotheses for both the biological and cultural evolution of language.

On the biological side, computer simulations might be used, for instance, to determine whether there is a strict correlation between the neural control necessary for the imitation of actions in apes and the neural activity involved in vocal imitation in humans [6]. As for the cultural evolution of language, simulations might be used to demonstrate the development of a communication system from scratch, or the cumulative cultural evolution of language over generations [7]. Of course, if we want to understand the whole picture, we need to explore how these aspects interact. It is clearly a task for the future to design experiments that integrate these aspects, and this will require better quantification of data and formalisation of theories.

1.2 Putting Order Into Language

In order to formalise theories of language evolution, we first need to formalise language itself. After all, we cannot suggest or explain mechanisms for something unless it has first been described. It is the job of linguists to put order in language. In linguistics, language can be viewed from two different but equally useful perspectives. Firstly, we can look at language synchronically, i.e. describe language at a single point in time, and concentrate on its systems and organisation. In contrast, the diachronic approach considers how languages change and diversify over time.

Using the **synchronic approach**, we can start off by breaking down language into its basic components. We might study the physical sounds of language, e.g. b, p, a, i, d, t (the subject of phonetics) and the ways they are organised to form meaningful elements, e.g. *bat*, *pat*, *bit*, *pit*, *bad*, *bid*, etc. (the subject of phonology). We might also analyse the structure of words and see how word roots (e.g. *clock*, *teach*) are combined with affixes to create grammatical inflections of the word (*clocks*, *teaches*) or new words (*clockwise*, *teacher*). This is the subject of morphology. Then we might look at the rules underlying how we combine words to form sentences (syntax). Together, the rules of phonology, morphology and syntax form what linguists call the **grammar** of a language. With the word "grammar", we are not referring to the set of rules prescribed by grammarians, such as "never split infinitives", but rather to the subconscious set of knowledge speakers have about all the categories of language (phonological, morphological and syntactic) and the rules by which they interact. Speakers of the same language may diverge in their pronunciation, word choice or morphology and syntax, but such **linguistic variation** (the subject of sociolinguistics) does not equate to language ungrammaticality. It does, however, reflect the fact that language is a dynamic system and, as such, is liable to change.

Synchronic approach: Approach in linguistics that considers language at a single point in time, often the present day, in order to understand the system behind it.

Grammar: In linguistics, the subconscious set of knowledge speakers have about the categories of language (phonological, morphological and syntactic) and the rules by which they interact. Sometimes refers just to the rules of syntax.

Linguistic variation: Different ways of saying the same thing. Variation occurs across all levels of grammar (phonological, syntactic, etc.) and can be attributed to physiological, psychological and sociological factors.

In the **diachronic approach**, we would look precisely at how languages have changed and developed over historical time. On the shortest timescale, language is in a constant state of flux due to the simple fact that there are different ways of saying the same thing. Inevitably, some trends catch on at the expense of others, leading to **language change**. Over a longer time span, language change leads to new varieties and eventually new dialects and new languages. Diachronic or historical linguistics studies the historical development of languages and attempts to develop general theories about how and why languages change. Comparative historical linguistics looks specifically at language families (e.g. Germanic, Italic, Balto-Slavic, Uralic, Dravidic) for evidence of yet older ancestral languages, or **proto-languages** (e.g. Proto-Indo-European, Proto-Uralic, Proto-Dravidian), but these date back at most 10,000 years to a time when the human language faculty was long up and running.

Diachronic approach: Approach in linguistics that considers the development of language over time.

Language change: Modification of language forms over time. The subject of interest in several subfields of linguistics: historical linguistics, sociolinguistics and evolutionary linguistics.

> **Proto-languages**: Postulated ancestral languages, e.g. Proto-Indo-European, from which attested languages are believed to have descended.

Today, through precise linguistic reconstruction, some are trying to identify a hypothetical common ancestor of all the world's languages (Proto-Human), supposedly going back 30,000 years or more. According to Ljiljana Progovac, Professor of Linguistics at Wayne State University in Detroit, linguistic reconstruction can contribute to the larger picture of language evolution by negotiating with the fields of neuroscience and genetics. She notes, "While linguistic reconstructions can identify ancestral proto-structures and distinguish them from more recent structures, neuroscience can test if these distinctions are correlated with a different degree and distribution of brain activation, and genetics can shed light on the role of some specific genes in making necessary connections in the brain possible" [8].

On an even larger timescale, evolutionary linguistics aims to explain the emergence of language from a non-linguistic system. Through insights from neuroscience and genetics, there is more and more recognition that the core features of language can be found in the pre-linguistic animal world. Prominent biolinguist Cedric Boeckx at the University of Barcelona started his career in the Chomskyan tradition of theorising about abstract principles "preset" in the human mind underpinning the structure of language. He now advocates, "If we believe that the target of linguistic theorising is 'ultimately biology', there is no alternative to going there, and doing some actual biology" [9]. He argues for a new kind of "biolinguistics enterprise" which steers away from the concepts of theoretical linguistics and focuses on data generated by biologists. He believes that linguistic structure is ultimately the result of "a collection of cognitive biases", and so rather than speculating about proto-languages and proto-structures, we should be building upon observed facts from the natural world.

In the next section, I will introduce the various interdisciplinary fields and sources of evidence that we can draw on to explain the origins of language.

1.3 Gathering the Evidence

1.3.1 Precursors in the Animal World

Understanding the origins and evolution of human language has been a longstanding question in evolutionary biology. By comparing human language with animal communication systems, researchers seek to identify commonalities and differences, which can shed light on the evolutionary pathways that led to the development of language in humans. In the last fifty years or so, findings have shown that animals communicate in a far more complex way than was previously assumed. For example, monkeys can produce differentiated calls to signal different enemies, and even combinations of calls to signal different situations. These simpler forms of communication can give clues as to the earliest form of human language.

The modern **comparative approach** goes beyond looking at communication signals and seeks precursors to human language in animal cognition. According to the cognitive view of language, linguistic structures reflect our mental representations of the world around us and the way they are processed is not specific to language but closely related to general cognitive abilities such as categorisation, perception, memory and reasoning. Some claim that every system in language evolved from some pre-existing cognitive capacity; for example, words may have arisen from the mental capacity to conceptualise objects and events, and syntax from the mental planning of actions. We shall delve into this more in Chapter 9.

Comparative approach: Comparative study of (i) structural features across different languages to highlight commonalities or diversity, or (ii) cognitive abilities across species with a view to understanding the evolutionary path to human language.

From the emerging comparative data, it is clear that many of the cognitive mechanisms needed to produce language are present in apes and were therefore present in our last common ancestor with apes. However, while many researchers agree that a host of animals can understand the meaning of individual words, and that some may understand simple sentences and syntactic variations, there is little evidence to suggest that any animal can manipulate signs to produce new sentences.

1.3.2   Infants' Language Learning

The emergence of language occurs not only on an evolutionary timescale but also during children's **language acquisition**. Babies seem to follow the evolutionary stages in language development when they learn language. In the first few weeks, their vocal sounds (mainly crying) simply reflect their physical and emotional state. The first coordinated movements of the vocal tract occur between two and four months when the first syllables are pronounced. During this period there is a lot of tongue thrusting and lip movement, which is thought to be a form of imitation learning. From four months onwards, all the different sounds start emerging, followed by experimentation with rhythm and intonation, which are used to signal not only emotions but also intentions. Before first words are produced, babies can already attribute meaning to utterances [10].

> **Language acquisition**: Process by which humans learn to use language. Requires exposure to the system and the use of innate capacities such as categorisation, imitation, memory and problem-solving.

Since we are not born with language in a ready-to-use state but have to learn it from other speakers, some have argued that language is primarily a social construct: the form of language depends on the arbitrary conventions agreed upon by the social group we live in. Since sounds and words are for the most part arbitrary, bearing no physical resemblance to the objects they refer to, they need to be learnt using the **memory** mechanism, which requires brain matter and time. It has been suggested that as early humans were confronted with new environmental conditions, a large brain able to process new information would have been a huge asset. As humans evolved ever larger and more complex brains, it became advantageous for babies to be born earlier, while their heads were still small. The continued rapid growth and development of the head and brain after birth encouraged language learning, which conveniently coincided with general learning about the world. This dual learning would have facilitated the crucial process of linking arbitrary linguistic symbols to meaning. We shall see what else children's language acquisition can reveal about the origins of language in Chapter 8.

> **Memory**: Cognitive system in which data or information is perceived, stored and retrieved when needed.

1.3.3   Semiotics

Semiotics, the study of the use of signs and symbols to communicate meaning, can help researchers explore the origins of human language in a number of ways. Firstly, semiotics intersects with biology and neuroscience in examining how the human brain processes signs and symbols. Investigating the neurological basis of symbolic thinking and communication can provide insights into how the cognitive mechanisms that underpin language evolved over time. Drawing parallels between the communication systems of humans and non-humans can shed further light on the evolutionary paths that may have led to the development of linguistic abilities in humans. In addition, semiotics considers the cultural and social aspects of signs and symbols, providing evidence of how languages adapt over time within societies.

> **Semiotics**: Study of the use of signs (linguistic and non-linguistic) to communicate meaning. Signs may imitate the objects they refer to (icons) or represent them in a conventionalised form (symbols).

Many, for example, Deacon (1997), have argued that the emergence of human language coincides with the emergence of complex symbolic communication. Animals are largely restricted to **symptomatic** signals, which are genetically determined responses to specific stimuli that can only convey information about immediate circumstances, such as danger or food availability. These signals are often **iconic** in nature, i.e. there is an intrinsic link between the signal and the message, as in growling to ward off an enemy. In human language, however, there is usually no obvious link between the sign and the concept—the word *table* has no physical resemblance to the object it refers to in the real world. In this case, the symbol used is **arbitrary**, determined by social convention, and therefore has to be learnt. Although chimpanzees can be taught to use arbitrary symbols, the resulting communication features basic one-to-one mapping between symbols and objects. In contrast, humans seem to be the only species to be able to handle a complex system of signs that interrelate with one another.

> **Symptomatic signalling**: Signals resulting from an internal state or emotion of the sender. Often consists of fixed, instinctual responses to specific stimuli.

> **Iconicity**: Existence of a natural connection between a sign's meaning and its sound or form.

> **Arbitrariness**: Absence of any natural connection between a sign's meaning and its sound or form.

However, human language is not simply a system of arbitrary symbols and rules but also makes use of iconicity: speakers use a multitude of facial expressions and manual gestures when they converse, and some words bear an acoustic resemblance to the concepts they refer to, for example, *slide*, *slip* and *slither*. Moreover, iconicity has been demonstrated to be especially prevalent in babies' first words, suggesting that it provides initial support in learning to map linguistic labels to concepts. It has long been suggested that iconicity might have played an important role in connecting sound with meaning in humans' first words. More recently, it has been argued that iconicity was also crucial in the emergence of grammatical structure, which is ultimately based on the ability to see "relational" analogy between form and meaning. For example, the way we sequence phrases in sentences usually mimics the real-world order of the events we are describing. Up until recently, the role of iconicity has been largely ignored in theories of language evolution. Some evolutionary scenarios adhere to the simplistic route of iconicity leading to arbitrariness, but it can be argued iconicity sits alongside arbitrariness as a fundamental property of language.

1.3.4 The Impact of Neuroscience

In recent years, neuroscientists have contributed to our knowledge of how language developed in humans by using neuroimaging to indicate what kind of brain structure and neural pathways are required for language. It is well known that Broca's area, positioned in the frontal lobe of the left hemisphere (in right-handed individuals), is responsible for the syntax (structure) of language. If this area of the brain is damaged, it leads to Broca's aphasia, where patients are unable to speak coherently, but are able to convey some sense of meaning in their utterances. It is very usual for prepositions, articles and linking words to be lost, so that "I'm going to the shops" may be rendered as "Go shop". Wernicke's area, on the other hand, situated in the left temporal lobe, deals with the semantics (meaning) of language. Patients with damage in this area of the brain are capable of forming grammatically correct sentences that sound fluent, but which contain no meaning—a condition known as Wernicke's aphasia. The following response would be typical for a sufferer of Wernicke's aphasia:

> A: *Hi, how are you today?*

> B: *I'm tall, are you pretty at home?*

Producing and understanding speech is an interconnected process that involves the coordination of these and multiple other brain regions. We shall learn more about speech areas of the brain and speech processing in Chapters 4 and 7.

1.3.5 Advances in Genetics

Thanks to recent developments in molecular techniques, our understanding of genetic variations between individuals has greatly expanded. It has long been recognised that certain speech and language disorders tend to run in families and have a strong hereditary component. By studying DNA samples from families affected by these disorders, geneticists have been able to identify specific genetic disruptions that are responsible for these conditions. The most frequently cited example of such a disruption concerns the **FOXP2 gene**, leading to impairments in expressive and receptive language abilities. According to British geneticist and director of the Max Planck Institute for Psycholinguistics Simon Fisher, the identification of genes like FOXP2 provides valuable insights into both the neurological foundations and the evolutionary origins of speech and language [11].

FOXP2 gene: Highly conserved gene in mammals. It is theorised that changes in the human version of this gene were essential for vocal control and thus language.

By comparing the DNA sequence of a known language-related gene with similar sequences found in other animal species, we can reconstruct the probable evolutionary history of the gene. We can determine when the gene first appeared and the alterations it underwent in different lineages, including our most recent ancestors. In the human version of FOXP2, two amino-acid coding changes have occurred since our evolutionary departure from chimpanzees. This discovery led to the hypothesis that these evolutionary changes might have played a crucial role in the emergence of speech and language in humans [11]. With the availability of comprehensive genomic data from different branches of the hominin family tree, it should be possible to empirically determine a timeline for other major steps towards language in our lineage.

1.3.6   Out of Africa

Leaving the "how" and "why" aside for the moment, let's turn to the "when" and "where". It stands to reason that human language emerged with human beings, but when did humans emerge? This is not such an easy question to answer. Roughly six million years ago, Hominini (a branch of the great apes in Africa) had differentiated into two branches: Panina (the ancestors of chimps and bonobos) and Hominina, commonly referred to as hominins (the ancestors of Australopithecus and Homo, or archaic humans). While original divergence between Panina and Hominina populations may have occurred as early as thirteen million years ago (in the Miocene), the earliest fossils clearly in the human and not the chimpanzee lineage date to around four million years ago (in the Pliocene), with "Australopithecus anamensis". Figure 1.1 shows an approximate timeline for the various branches that have emerged since the great apes.

As early Homo moved away from Africa, various species evolved. Roughly two million years ago, Homo erectus was thriving in East Asia. Other species established themselves in Europe, including Homo heidelbergensis (roughly 600,000 years ago) and Homo neanderthalensis (roughly 400,000 years ago). As these human species were evolving in Europe and Asia, evolution in East Africa did not cease. Roughly 300,000 years ago, Homo sapiens emerged, spreading to Asia and Europe in several waves, most of which seem to have ended in extinction. Present-day humans are

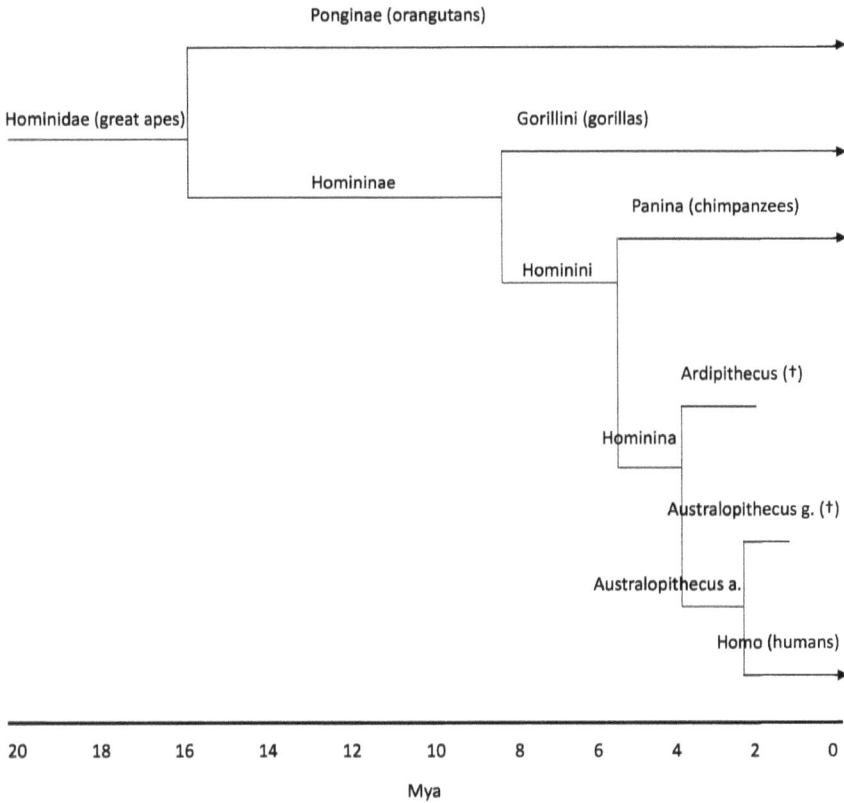

Fig. 1.1 The emergence of Homo

thought to have descended from a "recent" dispersal that took place roughly 70,000–50,000 years ago.

These new humans were much more sophisticated and, in all likelihood, had advanced language. Evidence of this comes from the field of **archaeology** and the discovery of elaborate tools, artefacts and cultural landscapes. Archaeological findings suggest that some groups were already living in the Middle East by 50,000 BP. From there, some spread westwards across Europe. Remnants of their culture dating from around 40,000 BP have been found in Spain and from around 35,000 BP in southwest France. Other humans spread eastwards, inhabiting northeastern Asia, eventually crossing from Siberia to Alaska and entering the American continent approximately 30,000 years ago [12]. All these successive migrations would have brought new varieties of language to different parts of the world.

Archaeology: Study of human history and pre-history via material culture.

Further evidence for the origins of language comes from **palaeontology**. The discovery in 1989 of a Neanderthal hyoid bone (the horseshoe-shaped bone "floating" in the neck at the base of the jaw) and the fact that it closely resembled the human hyoid bone (see Fig. 1.2) suggests that Neanderthals were anatomically capable of producing sounds similar to those of modern humans. In humans, the hyoid bone is supported by a myriad of finely tuned muscles that link up in all directions, including to the tongue (see Fig. 1.3). It allows a wider range of tongue movements and therefore plays a vital role in speech.

Palaeontology: Study, through fossil remains, of life prior to the Holocene (the current geological epoch starting roughly 12,000 years ago).

Research shows that Neanderthals also had a hearing range necessary to process human speech. The 2006 analysis of the Neanderthal **genome** revealed that they possessed a FOXP2 gene similar to that of humans, though probably not identical.

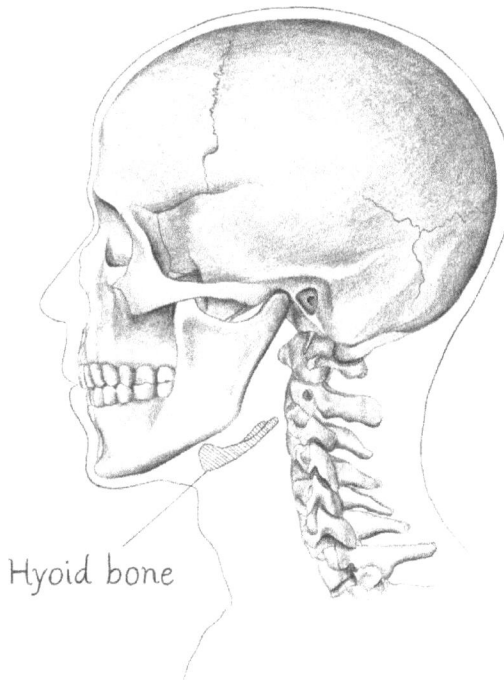

Hyoid bone

Fig. 1.2 Position of the hyoid bone (by William Duffy)

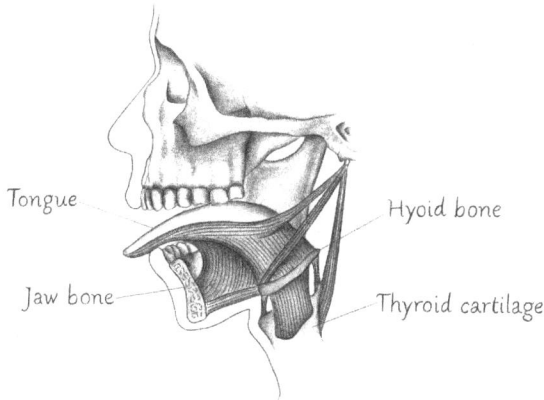

Fig. 1.3 Muscular attachments of the hyoid bone (by William Duffy)

Neurologically, Neanderthals had an expanded Broca's area, pointing to syntax. The degree of language complexity is difficult to establish but given that Neanderthals achieved some technical and cultural complexity, and interbred with humans, it is reasonable to assume they were at least fairly articulate [13].

> **Genome**: Complete set of genetic information in an organism. Provides all of the information an organism requires to function.

Some place the development of primitive language-like systems (protolanguage) with the appearance of Homo habilis, some 2.5 million years ago, while others place it only with Homo erectus (1.8 million years ago) or with Homo heidelbergensis (0.6 million years ago). It seems likely that when Homo sapiens migrated to Europe some 70,000 years ago, they were carrying some form of language with them. This is supported by evidence of symbolic behaviour, such as the use of tools, art and burial rituals, which suggest the presence of complex cognitive abilities and communication skills. Neanderthals died out roughly 30,000 years ago, making us, Homo sapiens, the only surviving human species on earth. How did we manage to settle so rapidly in so many habitats and push all other human species into oblivion? Some researchers believe that the ability to speak a complex, highly productive and versatile language was a crucial factor in the success of Homo sapiens as they migrated across the globe and adapted to new environments.

1.4   Chapter Summary

We started with a brief overview of the historical development of the discipline of evolutionary linguistics, which has been characterised by various key figures and movements. An age-old interest in the origin of language was reignited when Charles Darwin applied his general theory of evolution to human beings and assumed that language itself may also have evolved through natural selection. He suggested that language started as musical cadences for courtship that became more and more elaborate through competition. The behaviourists, inspired by B. F. Skinner, claimed that language was rather a set of habits learnt from the environment. This idea was overturned by the nativists, who argued the language was too complicated to be picked up by imitation. The nativists split into two factions: the saltationists (led by Noam Chomsky), who believe that language is the product of a single genetic endowment in humans alone, and the adaptationists (championed by Chomsky's student, Ray Jackendoff), who believe that language is a complex adaptation that evolved in stages. In the twenty-first century, the culturalists have been claiming language can be more plausibly explained as the product of cultural evolution. The modern approach is to recognise and integrate both the biological and cultural evolution of language. For this to happen, there needs to be better quantification of data and formalisation of theories.

Next, we dealt with the formalisation of language (putting order into language). Using the synchronic approach, we can break down present-day language into its basic components (sounds, words and sentences) and learn about its structure and organisation. Diachronic or historical linguistics examines how language changes and develops over time. Comparative historical linguistics attempts to find links between known languages with a view to reconstructing possible earlier common ancestral languages. On an even larger timescale, evolutionary linguistics aims to explain the emergence of language from a non-linguistic system.

Finally, we looked at the various branches of knowledge and sources of evidence required to explain the origin of language, including archaeology, palaeontology, genetics and neuroscience, as well as research on animal cognition and children's language learning. Each of these areas will reappear throughout the book, but our next task is to describe our object of investigation and look more closely at the design of language. As we shall see later in the book, a description of the various systems of language is essential in understanding what evolved, why it evolved the way it did and how it changed over time.

References

1. Jackendoff, R. (1999). Some possible stages in the evolution of the language capacity. *Trends in Cognitive Sciences 3*, 272–79.
2. Darwin, C. (1871). *The descent of man*. John Murray.
3. Deutscher, G. (2005). *The unfolding of language*. Metropolitan Books.

4. Nölle, J. (2014). A co-evolved continuum of language, culture and cognition: Prospects of interdisciplinary research. *Studies about Languages,* 2014 No. 25. https://doi.org/10.5755/j01.sal.0.25.8504
5. Zuidema, W. (2005). *The major transitions in the evolution of language.* University of Edinburgh
6. Cangelosi, A., & Parisi, D. (2001). Computer simulation: A new scientific approach to the study of language evolution. In A. Cangelosi & D. Parisi (Eds.) *Simulating the evolution of language.* Springer.
7. Nölle, J., Hartmann, S., & Tinits, P. (2020). Language evolution research in the year 2020. *Language Dynamics and Change, 10,* 3–26. https://doi.org/10.1163/22105832-bja10005
8. Progovac, L. (2016). A gradualist scenario for language evolution: Precise linguistic reconstruction of early human (and Neandertal) grammars. *Frontiers in Psychology, 7,* Article 1714. https://doi.org/10.3389/fpsyg.2016.01714
9. Boeckx, C. (2021). Reflections on language evolution: From minimalism to pluralism. In M. Dingemanse & N. J. Enfield (Eds.), *Conceptual foundations of language science 6.* Language Science Press.
10. Crystal, D. (1987). *The Cambridge encyclopedia of language.* Guild Publishing.
11. Fisher, S. E. (2017). Evolution of language: Lessons from the genome. *Psychonomic Bulletin & Review, 24*(1), 34–40. https://doi.org/10.3758/s13423-016-1112-8
12. Aitchison, J. (2000). *The seeds of speech.* Cambridge University Press.
13. Wikipedia. (2023). *Neanderthal.* https://en.wikipedia.org/wiki?curid=27298083#Language

Chapter 2
The Design of Language

Language is a defining feature of humanity. It is a highly structured system of communication that allows humans to convey an endless range of messages using a finite set of signals. These signals may be sound-based (as in speech), visual (as in writing or sign language) or tactile (as in braille). Within the synchronic framework for analysing language, which describes language at a single point in time, there are two distinct approaches, both of which are relevant to language evolution studies. The first, the **structuralist approach**, developed by Swiss linguist Ferdinand de Saussure in his seminal *Cours de linguistique générale* (1916), aims to break language down into its elements (i.e. sounds, words, sentences, etc.) and explain how these elements are organised and relate to one another. It emphasises language as a closed system and in this respect is related to nativism. The second, the functionalist approach, which spawned in the 1920s to 1930s out of Saussure's systematic structuralist approach, seeks to identify what language is used for. In this chapter, we will focus on the first method of analysing language.

> **Structuralist approach**: Method in linguistics that aims to break language down into its elements (sounds, words, sentences) and explain how these are organised and relate to one another. Language is conceived as a self-contained system of interconnected units.

At this point, it should be mentioned that in a discussion about the origins of language, we often favour the spoken form (i.e. speech). Nevertheless, in recent decades, there has been a shift to viewing language as inherently multimodal, involving speech and gesture, with some theories of language origins suggesting that gestural communication was a forerunner to speech. This means that **sign languages** have become vital to language evolution research, especially those that emerge spontaneously in communities without a pre-existing conventional sign system, since

J. Dornbierer-Stuart, *The Origins of Language*,
https://doi.org/10.1007/978-3-031-54938-0_2

these shed light on the natural processes involved in the emergence of language. We will learn more about one such sign system in Chapter 5.

> **Sign language**: Language that uses the visual-gestural modality to convey meaning. Sign languages are considered fully fledged natural languages with their own grammar and lexicon.

2.1   An Ingenious Design

Human language is a brilliantly engineered tool. On the one hand, we have a finite number of elements (speech sounds, e.g. /p/, /t/, /k/) which on their own are meaningless. On the other hand, these sounds are combined in specific ways to produce a large number of meaningful elements (words), which in turn can be combined following particular rules to form an infinite number of sentences. In 1960, the American linguist and anthropologist Charles Hockett suggested that this feature, which he called **duality of patterning**, evolved when a growing number of concepts needed to be expressed. Researchers today often agree it is a "cultural" phenomenon, arising in response to the pressures placed on language as it is used communicatively.

> **Duality of patterning**: Feature of language whereby meaningless elements (speech sounds) are combined into meaningful elements (words), which are combined further into phrases and sentences.

However, the ability to combine elements into new elements must have first required the biological evolution of mechanisms to deal with **combinatorial structure**. Language researchers Zuidema and de Boer [1] suggest that many of the biological prerequisites for combinatorial structure (for example, the ability to recognise and categorise different sounds) have been inherited from our primate ancestors and are based on ancient cognitive mechanisms to detect structure in the environment. Nevertheless, they concede that combinatorial structure will have also evolved from pressure to increase the number of signals brought about by the increased need to communicate different meanings.

> **Combinatorial structure**: Feature of language whereby smaller elements, such as sounds or words, are combined to form larger elements, such as words or sentences.

Returning to the design of language, we have seen that language can be broken down into different types of elements (sounds, words, etc.). These elements are all part of a **hierarchical structure**, which is headed by the sentence or clause. For example, in the clause "The girls in the house were singing lullabies", the word "girls" is embedded in the noun phrase "The girls", which, in turn, is embedded in the larger phrase "The girls in the house", which, in turn, is embedded in the clause "The girls in the house were singing lullabies". This can be represented mathematically as follows:

[[[The [girls]] in the house] were singing lullabies]

When we formulate sentences, we cannot jiggle words around willy-nilly but must adhere strictly to this underlying phrasal structure, and this is a feature of all languages.

Hierarchical structure: Organising principle in language whereby smaller elements (e.g. words) are embedded into larger elements (e.g. phrases), which are in turn embedded into yet larger elements (e.g. clauses and sentences).

In addition to the hierarchy of syntax (which puts order to words and phrases in a sentence), words themselves have an internal hierarchical structure. The word "girls" consists of two morphemes: a root (GIRL), bearing the core meaning of the word, and an inflectional ending (-S), which adds grammatical meaning to the word. Each morpheme can be divided further into phonemes (speech sounds). The morpheme GIRL consists of three distinctive sounds in Standard British English: /g/, /ɜ:/ and /l/.

On another level, the musical system running through speech (known as prosody) also has its own hierarchical structure. At the top of this hierarchy is the utterance (similar to sentence), followed by the intonational phrase—a chunk of speech typically carried out in one breath and centred around a prominent word (roughly corresponding to a grammatical clause):

/ the girls in the house were singing lullabies /

The intonational phrase can be broken down into phonological phrases, again each containing a prominent word (roughly corresponding to a grammatical phrase):

[the girls] [in the house] [were singing] [lullabies]

Each phonological word can be broken down into syllables—groups of sounds pronounced in a single beat (roughly corresponding to morphemes). "Singing" has two syllables:

sing-ing

Although these phonological chunks align fairly well with syntactic structure, they are actually processed in different parts of the brain and in different ways. MRI studies have shown that the right hemisphere of the brain deals with prosodic processing and the left hemisphere deals with syntactic and semantic processing, but the exact timing of each type of processing and their interaction is still debated. As we shall see later in the book, different language structures require different processing strategies in the brain. On the one hand, we use prosodic cues to break down a continuous stream of speech into meaningful elements as it arrives in real time. On the other hand, we fit these elements into a preconceived grammatical structure that has been learned and stored in the memory. What is clear is that language is not simply a linear sequence of sounds but a highly organised collection of hierarchical systems which are superimposed on one another and intricately interwoven.

2.2   The Sound System (Phonology)

We will now look at the smallest discrete units (or **segments**) of sound and see how these combine to form words. Each language has its own set of distinctive sounds, or **phonemes**, that are used to distinguish one word from another: /p/ and /t/ are two separate phonemes since "pin" has a different meaning from "tin". English has roughly 43 distinctive sounds, not to be confused with letters of the alphabet, of which there are only 26. A single letter of the alphabet can represent more than one sound in English. For example, the letter "c" can be pronounced /k/ as in *cat* or /s/ as in *receive*. Likewise, the letters "th" can be pronounced /θ/ as in *think* or /ð/ as in *this*. For each distinctive speech sound, however, there is only one corresponding phonetic symbol.

Phonology: Component of grammar that deals with how sound is organised to convey meaning. Includes the elementary units of sound (phonemes) and how these combine, as well as features that accompany speech sounds (e.g. stress and rhythm).

Segment: Any discrete unit that can be identified auditorily in a stream of speech, e.g. phoneme, syllable or word.

Phoneme: Abstract unit of sound that can distinguish one word from another in a particular language, e.g. /p/ and /t/ are phonemes in English since "pin" has a different meaning to "tin". Phonemes are usually written between slashes / /.

So, how do speech sounds distinguish themselves? Let us first see what they have in common. Producing a speech sound involves expelling an airstream from the lungs through the mouth and, at the same time, modulating it in a particular way to produce the desired sound. All phonemes share common features. In English, just over half are consonants, which are produced with a complete or partial closure of the vocal tract, e.g. /p/, /d/ and /w/. The rest are vowels, which are produced without obstruction of the vocal tract, e.g. /æ/ as in *pat* and /ɑː/ as in *father*. All vowels are voiced, which means they are pronounced with a vibration of the vocal cords. Consonants can be either voiced, as with /b/, /d/ and /g/, or voiceless, as with /p/, /t/ and /k/ (compare voiced /b/ in "bin" and voiceless /p/ in "pin").

In addition to the voiced-voiceless distinction, consonants can vary in their manner of articulation (how air escapes from the vocal tract). /p/, /t/, /k/, /b/, /d/ and /g/ are stops (explosive), /f/, /v/, /s/ and /z/ are fricatives (turbulent), /m/ and /n/ are nasals (air escapes through the nose) and /l/, /r/, /j/ and /w/ are approximants (these are produced by narrowing but not blocking the vocal tract). Consonants vary further in their place of articulation (where in the vocal tract the narrowing or obstruction occurs). For example, /p/, /b/ and /m/ are bilabial (both lips come together), /t/, /d/ and /n/ are alveolar (the tip of the tongue is placed against the gum ridge) and /k/ and /g/ are velar (the back of the tongue is pressed against the soft palate).

Table 2.1 shows the 23 distinctive consonant sounds found in modern Standard British English. The symbols are taken from the International Phonetic Alphabet (IPA), which lists the majority of phonemes found in all human languages (well over 100 distinct consonants and vowels).

Vowels have no obstruction but can vary according to the position of the tongue. This can be placed higher in the mouth (say *eeh*) or lower (say *aah*) and towards the front of the mouth (say *geese*) or the back (say *goose*). Vowels can also vary in their length. For example, /æ/ and /ɪ/, as in *pat* and *pit*, are short, whereas /ɑː/ and /iː/, as in *father* and *feet*, are long.

Table 2.1 Consonant sounds used in modern Standard British English

	Nasal	Plosive		Fricative		Approximant
Labial	m	p	b	f	v	w
Dental				θ *think*	ð *this*	
Alveolar	n	t	d	s	z	l
Post-alveolar				ʃ *ship*	ʒ *vision*	r
Palatal						j *you*
Velar	ŋ *sing*	k	g	x *loch*		
Glottal				h		

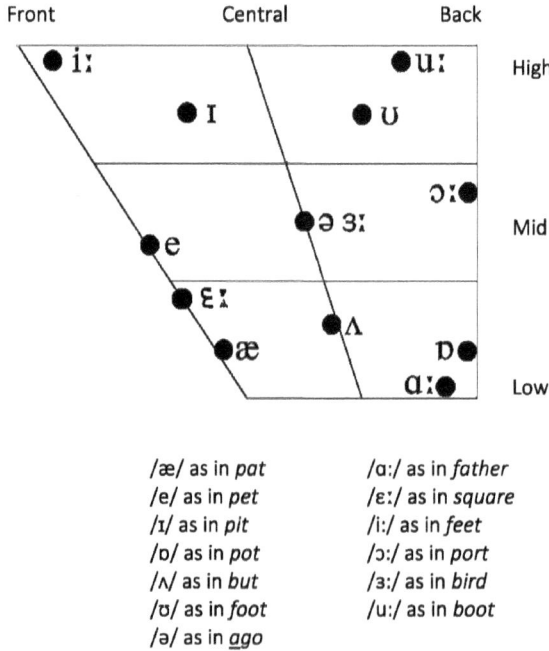

Front	Central	Back	
● iː		● uː	High
● ɪ		● ʊ	
		ɔː●	
	●ə ɜː		Mid
●e			
●ɛː			
●æ	●ʌ	ɒ●	
		ɑː●	Low

/æ/ as in *pat*   /ɑ:/ as in *father*
/e/ as in *pet*   /ɛ:/ as in *square*
/ɪ/ as in *pit*   /i:/ as in *feet*
/ɒ/ as in *pot*   /ɔ:/ as in *port*
/ʌ/ as in *but*   /ɜ:/ as in *bird*
/ʊ/ as in *foot*   /u:/ as in *boot*
/ə/ as in *ago*

Fig. 2.1 Vowel sounds used in modern Standard British English

Figure 2.1 shows the thirteen distinctive vowel sounds used in modern Standard British English. The vertical axis denotes the height of the tongue in the mouth and the horizontal axis denotes how far forward the tongue is located.

The description of sounds becomes a little more complicated because each phoneme can have variant sounds, or **allophones**, according to the other sounds around it. For example, English speakers pronounce the p in *pin* differently from the p in *spin*. The p in *pin* is breathier than the p in *spin*. The former is said to be aspirated (= [pʰ]), and the latter unaspirated (= [p]). There is a **phonological rule** that says that a voiceless stop such as /p/ is aspirated when it occurs at the beginning of a word (pin = [pʰɪn]), but when it occurs after /s/, it is unaspirated (spin = [spɪn]). These slight sound changes usually reflect the need to facilitate awkward articulations. Substituting one allophone with another (e.g. [pʰ] with [p]) does not affect the meaning of a word, whereas substituting one phoneme with another (e.g. /p/ with /t/) creates a new word. Humans are particularly good at only paying attention to meaningful distinctions.

Allophones: Sound variants of a single phoneme that do not produce a change in word meaning. Some allophones ease awkward articulations, e.g. aspirated t in *top* becomes unaspirated t in *stop*; others are dialect-dependent, e.g. intervocalic t in *water*,

> pronounced [t] in Standard British English, becomes flapped t (= [ɾ]) in American English. Allophones are usually written in square brackets [ ].

> **Phonological rules**: Rules underlying how we pronounce phonemes in different phonological environments, either within words (e.g. [n] in *thorn* assimilates to [m] in *Thornberry*) or between words (e.g. [d] in *mad* assimilates to [g] in *mad goat*).

Having identified what phonemes are, we now need to see how phonemes combine to form words. Every language has certain permitted sequences of sounds which are determined by the **phonotactic rules** of the language. In English, for example, when three consonants start a word, as in *spring*, *string* and *spleen*, the first sound can only be /s/, the second can only be /p/, /t/ or /k/ and the third can only be /l/, /r/, /w/ or /j/. Czech allows up to four consonants at the beginning of a word, again following strict rules of progression.

> **Phonotactic rules**: Rules underlying how we configure sequences of sounds. For example, in English, if a word begins with /s/ and the following sound is a stop, that stop must be voiceless. The rules are highly variable between different languages.

As we saw in Sect. 2.1, speech sounds are organised into rhythmic beats, or syllables. Syllables can be broken down further into an onset, a nucleus (usually a vowel) and a coda, e.g.

Onset	Nucleus	Coda
s	i	ng
spr	i	nt

The permitted sequence of sounds in onsets is different from that in codas. In English, the sound sequence "spr" only works at the beginning of a syllable and "nt" only works at the end. This actually helps us recognise where words begin and end in a constant stream of speech. Most languages prefer consonants in the onset rather than in the coda. In fact, the Polynesian language Tongan actually requires consonants in the onset and prohibits them in the coda, meaning it has only one syllable type: CV (consonant + vowel).

Thus, how we configure sounds depends on both innate biological constraints (such as ease of articulation) and cultural preferences. We will now look more closely at the meaningful elements created through the combining of speech sounds, i.e. words and morphemes.

2.3   The Word System (Morphology)

Words are not simply sequences of sound but also carry with them meaning. A word, or **lexeme**, can be defined as the smallest sequence of sounds that can stand alone with objective meaning—*clock* means something but *cl* or *ck* does not. Words are made up of smaller meaningful elements, roots and affixes, known as **morphemes**. For example, the word *teaches* consists of the root TEACH, which can also stand alone, like a word, and the affix -ES, which has a purely grammatical meaning and cannot stand alone.

Morphology: Component of grammar that deals with words, their meaning and how they are structured.

Lexeme: Smallest sequence of sounds that can stand alone with objective meaning and be moved to different places (grammatical slots) in sentences while staying intact, e.g. *smile, wheelbarrows*.

Morpheme: Smallest meaningful part of a word, either a root or an affix. A root (e.g. CLOCK) generally refers to something existing in the outside world and may be unadjoined, whereas an affix (e.g. -S) only carries grammatical meaning and cannot stand alone.

As we have already seen, there is often no obvious link between the form of a word and its meaning—the word *table* has no physical resemblance to the object it refers to in the real world. Nevertheless, some words imitate the sounds associated with their referents (the objects or actions in the outside world they refer to), e.g. *hiss, buzz*. This is an example of **onomatopoeia**, a type of iconicity where the sounds in words are used to mimic the sounds of their referents. In addition, some researchers have found that words that are closely connected semantically have more closely aligned phonological forms than would be expected by chance. For instance, in English, a large number of words beginning with /sp/ pertain to water (e.g. *splash, splatter, spray, spill*). In another type of iconicity, the sounds in words are used to mimic not the sounds of their referents but other sensory properties or emotions related to them. For example, /fl/ is often used to refer to movement (e.g. *flip, flap, flicker*) and /sl/ to refer to something negative or repellent (e.g. *slime, slur, slum*).

> **Onomatopoeia**: Feature of language whereby a word representing an object or event imitates the sound associated with the object or event.

Some believe that humans' first naming words relied heavily on iconicity, before vocabularies grew too large and had to start making use of arbitrariness. A growth of vocabulary and arbitrariness would have no doubt driven the development of memory in humans. The power of the memory of modern humans cannot be underestimated. We have thousands and thousands of words stored in our brains, and each word itself is a store of information containing details on its phonological shape (how it is pronounced), all its nuances of meaning and how it is used in a sentence.

So far, we have described sounds, seen how they combine to form words and described some of the characteristics of words. We will now turn to how words combine to form phrases and sentences.

2.4 Grammar (Syntax)

In addition to referring to things in the world, words also carry grammatical information. Firstly, words belong to different grammatical categories, such as nouns, adjectives, verbs, prepositions, conjunctions, etc. Each category has its own grammatical properties; for example, nouns typically indicate number (singular, plural), gender (masculine, feminine) and case (nominative, accusative, etc.) and verbs often carry information about the time of an action (tense), the duration or completion of an action (aspect) and the possibility/probability of an action (mood). We shall see more about nouns and verbs and their foundations in mental thought in later chapters.

Secondly, each word imposes restrictions on other words according to its form and the **syntactic rules** of the language concerned. For example, the form of one item, e.g. the singular noun *duke*, forces a second item in the sentence, e.g. the verb *march*, to appear in a particular form, in this case, the singular (*marches, is marching*, etc.). This is known as agreement. Syntactic rules also determine how words or elements are ordered in a sentence. This happens on a number of levels. Firstly, words are combined with other words to form larger categories (phrases):

duke	→	*the old duke*
noun (N)		noun phrase (NP)
march	→	*is marching*
verb (V)		verb phrase (VP)

This usually happens in a particular order, depending on the language. In English, for example, adjectives precede nouns in noun phrases, whereas in French, adjectives follow nouns. These phrases then fill different grammatical slots in sentences, which linguists have labelled Subject (S), Verb (V), Object (O), Complement (C) and Adverbial (A). Thus, we have:

The old duke	*is marching*	*his men*	*to the top of the hill*
NP	VP	NP	PrepP
S	V	O	A

These grammatical slots also have a set order, depending on the language. English generally follows the subject-verb-object (SVO) order in its clauses, as in:

Peter	*ate*	*cake*
S	V	O

Finally, clauses can have other clauses attached to them or inserted inside them, e.g.

[Peter	*ate*	*cake]*	*and*	*[Tina*	*drank*	*tea]*
S	V	O		S	V	O

Peter	*[who*	*was*	*hungry]*	*ate*	*cake*
S	[S	V	C]	V	O

This means that there is no limit to the number of possible sentences that can be formed. A clause can always be attached to or inserted into another clause, ad infinitum, making language a powerful and boundless means of expression. We shall see more about the cognitive foundations of word order in later chapters.

> **Syntax**: Component of grammar that deals with how we combine words into phrases and sentences.

> **Syntactic rules**: Rules underlying how words are ordered (e.g. SVO) and constrained (e.g. by agreement) in sentences.

2.5 Meaning (Semantics)

Semantics, the component of grammar that deals with meaning, is linked to all other components of grammar. We have already seen that meaning can sometimes be conveyed through individual sounds (e.g. /sl/ is often used to refer to negative or repellent properties) and, of course, through words. Not only does a word have **referential meaning** (i.e. it refers to something in the world), but it is also related to other words in a web of semantic structure. For example, *dog* is a hyponym (subcategory) of *animal*, *arm* is a meronym (smaller part) of *body*, *large* is a synonym of *big* and *big* is an antonym of *small*. Through observing slips of the tongue (see Chapter 7), psycholinguists have demonstrated that words are stored in our **mental lexicon** not alphabetically, as in a dictionary, but rather on the basis of associations between semantically related lexical items. This is the reason we might mistakenly say "hot" instead of "cold", "arm" instead of "leg" or "itch" instead of scratch", the last two words being particularly "welded at the hip" semantically.

Semantics: Component of grammar that deals with how meaning is conveyed in a language.

Referential meaning: A word has referential meaning if it refers to something existing in the world, irrespective of the present situation. Also known as lexical meaning.

Mental lexicon: A person's "mental dictionary". Set of knowledge a speaker has about words and their meanings.

Meaning is also conveyed through syntax and word order. For example, *The boys chased the girls* has a different meaning to *The girls chased the boys*. In the first sentence, the position of *boys* before the verb tells us that the boys are the "doer" (or **agent**) of the action. The position of *girls* after the verb tells us that the girls are the "sufferer" (or **patient**) of the action. Thus, nouns and noun phrases, in addition to containing lexical meaning, correspond to **thematic relations** (i.e. they have semantic roles) within the sentence.

Agent: One of the semantic roles of a noun phrase within a sentence. The thing or person that does an action.

Patient: One of the semantic roles of a noun phrase within a sentence. The thing or person that undergoes an action.

> **Thematic relations**: The various semantic roles that a noun phrase can play within a sentence (e.g. agent, patient). At the interface of syntax and semantics.

On top of the meaning provided by speech sounds, words and syntax, more meaning is provided by the **prosody** of an utterance. Prosodic features, or **suprasegmentals**, are those elements of sound such as stress, rhythm and intonation that extend across syllables and larger units of speech. They can reflect the grammatical function of an utterance, e.g. whether it is a statement (*It's a cat*) or a question (*It's a cat?*), show emphasis or contrast (*This one, not that one*), or reveal the attitude of the speaker, e.g. *Really?* can be stressed differently, according to whether the speaker wishes to show interest, indifference, approval or disapproval. In the next chapter, we shall learn more about these features and how they scaffold language, and we will consider the implications for language origins.

> **Prosody**: System of rhythm, stress and intonation in speech. Provides important information beyond literal word meaning.

> **Suprasegmentals**: Features of speech, such as stress, rhythm and intonation, that are not phonetic segments (vowels and consonants) but accompany speech sounds, syllables and larger units of speech.

Finally, there are contextual features "outside" of the utterance (the subject of **pragmatics**) that can contribute to the meaning of an utterance. These include the verbal context (the language surrounding the utterance), the cultural context (what is understood due to cultural knowledge) and the situational context (what is clear from the setting and circumstances of the utterance). For example, the preceding verbal context is important for interpreting pronouns. For instance, in

John went to the store. He bought some groceries

we know that "He" refers to John. The cultural context can influence the understanding of idioms and expressions. For instance, the meaning of "kick the bucket" may not be evident without cultural knowledge of the expression. Finally, the situational context can be vital for understanding certain exchanges. For instance, the following interaction might seem quite bizarre:

A: *I have a fourteen-year-old son*
B: *Well, that's all right*
A: *I also have a dog*
B: *Oh, I'm sorry*

until we have the background information that A is raising a series of possible disqual-
ifications for apartment rental with the landlord B [2]. Meaning is thus contextual
and collaboratively made.

Pragmatics: Aspect of meaning that comes from the context of an utterance.

According to British evolutionary linguist and cognitive scientist Thom Scott-
Phillips [3], the academic discipline of linguistics has for most of its history been
concerned with the form and structure of language, with little regard for the connection between language and the outside world. Scott-Phillips believes that the neglect
of pragmatics in language evolution research is a profound mistake. He notes that
the way languages are used in communication is critical to any evolutionary explanation of the forms that languages take. He argues that language is a direct extension
of social intelligence, which enables humans to express and recognise intentions,
and which depends on a very sophisticated "theory of mind". This will be explored
further in the following chapter.

2.6   Chapter Summary

At the beginning of the chapter, we saw how language can be viewed as a limited
set of discrete sounds which, on their own, lack meaning but, when combined using
an array of phonological and syntactic rules, form words and sentences that convey
meaning. This very structuralist view of language gives us the necessary theoretical
framework on which to base our understanding of speakers' knowledge of grammar.
However, as we saw in our reflections on pragmatics, language is highly dependent on
the broader communicative context in which it is used, and thus involves much more
than a knowledge of grammar. It also requires knowledge of how to use language
successfully in everyday interactions with fellow human beings. This is why when
hypothesising about how language came to be the way it is, we need to think beyond
structure and look at the function of language. This is the subject of the next chapter.

References

1. Zuidema, W., & de Boer, B. (2018). The evolution of combinatorial structure in language. *Current Opinion in Behavioural Sciences, 21*, 138–144. https://doi.org/10.1016/j.cobeha.2018.04.011
2. Levinson, S. C. (1983). *Pragmatics*. CUP.
3. Scott-Phillips, T. C. (2014). *Speaking our minds*. Palgrave Macmillan.

Chapter 3
The Purpose of Language

As a complement to Saussure's structuralist approach to language, the **functionalist approach** considers the purpose of language and the context in which linguistic elements are used. It is a development from structuralism in that it extends the analysis of language from a focus on its internal structure to a broader consideration of its communicative functions and social context. This shift from an abstract, system-based perspective to a more contextual and pragmatic viewpoint can be seen to align with adaptationist theories of language evolution, which view language as evolving in response to environmental challenges and communicative needs. In this chapter, we will first investigate the many functions of modern language according to Geoffrey Leech's classification, and then link each function to some traditional theories of language origins as well as to some more recent hypotheses on how language emerged.

> **Functionalist approach**: Method in linguistics that seeks to explain language structure in relation to what language is used for.

Intuitively, we might assume that language is primarily used to convey useful information, such as "The train arriving at Platform 1 is the fast train to London", and thus a means of coordinating the actions of fellow human beings. But what about greeting, thanking, apologising, making requests, threats, promises or small talk? The purposes that language serves are so numerous that it is difficult to say what its original function was. Language is not always efficient when it comes to dealing with information; to explain complex concepts, you might need to use many words, which can be time-consuming and potentially confusing. Moreover, language can be ambiguous; some words or phrases have multiple meanings, which can lead to misunderstandings. In technical fields, ambiguity can lead to serious errors or confusion. It could therefore be argued that language is more geared to social interaction and maintaining social relationships. The question then arises whether language was

© The Author(s), under exclusive license to Springer Nature Switzerland AG 2024
J. Dornbierer-Stuart, *The Origins of Language*,
https://doi.org/10.1007/978-3-031-54938-0_3

"designed" for a social function, or whether it was "adapted" to complement our highly complex social organisation. We are not the only animals to have complex social structures, but only we have come up with complex language.

By studying the functions of contemporary language, we can gain insights into the communicative needs and social dynamics that may have influenced the development and evolution of language over time. In an attempt to systematise how language is used today, the influential British linguist Geoffrey Leech (1936–2014) introduced five functions of language: the expressive, directive, social, informational and aesthetic. We will now explore these functions and attempt to place each in the evolutionary context.

3.1 The Expressive Function

Let us begin with Leech's "expressive" function of language. This refers to the language we use to express feelings and emotions, which is for the most part involuntary and might therefore be compared to the symptomatic calls of animals (see 1.3.3). Examples of present-day expressive language include exclamations to express fear (*yikes!*), surprise (*wow!*), joy (*yay!*), pain (*ouch!*) or foreboding (*uh-oh!*). A popular theory of language origins, already held by some ancient Greek philosophers and revived by Rousseau in the eighteenth century, saw humans' first words as emotional interjections triggered by such emotions. Over time, these primal sounds could have evolved and become more structured, eventually developing into the complex linguistic systems we have today. The so-called "interjectional theory" was later harshly dismissed and disparagingly renamed the "pooh-pooh" theory by leading nineteenth-century philologist Friedrich Max Müller. It was assumed that the expressive sounds people make as an emotional release can hardly be considered the basis for the majority of words and are hence unlikely to be the source of language proper. Fig. 3.1 offers a humorous take on this discussion.

However, over a century later, certain researchers, e.g. Wray [1] and Arbib [2], have been revisiting the theory within the "**analytic model of language evolution**". This model suggests that humans started with expressive **holistic** vocalisations (vocalisations which refer to whole ideas or situations and which cannot be broken down into smaller meaningful parts) before slowly isolating individual chunks and associating these with meaning. In fact, it was Darwin who first suggested this. In *The Descent of Man* (1871), having noted "a widely-spread analogy" between language and birdsong, he went on to suggest, with his "musical origins" hypothesis, that humans first used holistic musical cadences for courtship and territorial defence and that these eventually transitioned to meaningful language [3]. However intuitive Darwin's model might have been, it failed to explain how musical cadences transitioned to meaningful language. It took linguist Otto Jespersen [4] to suggest that from this initial holistic stage, a cognitive process of analysis could have gradually isolated individual chunks of the musical phrase and associated them with

Fig. 3.1 The emergence of language (Source: CartoonStock Ltd.)

meaning [5]. Thus, the "analytic model" and the "musical origins" hypothesis can be considered to complement each other.

Analytic model of language evolution: Assumes humans started with holistic vocalisations and slowly isolated individual chunks and associated these with meaning.

Holistic: Concerned with wholes rather than separation into parts.

The "analytic model" contrasts with the "**synthetic model of language evolution**", advocated by, e.g. Bickerton [6], Jackendoff [7] and Tallerman [8], which assumes words evolved first, succeeded by syntactic operations to combine them. According to proponents of the "synthetic model", words could not have been picked out of holistic vocalisations since these are regulated by different areas of the brain to the language of intentional communication. Rather, it is more likely that nouns and verbs developed from cognition, along with the need to refer to more concepts and the possibility to produce more sounds (through enhanced vocal control). These contrasting views will be explored later in the book.

> **Synthetic model of language evolution**: Assumes language began with small elements of meaning (words) which were combined to form larger elements of meaning (sentences).

Other popular and perhaps unjustly ridiculed hypotheses of language origins were provided by the "natural sound source theories". The basic idea of these theories is that humans' first words were imitations of sounds from the natural environment. The "ding-dong" theory (attributed to Müller himself!) holds that all objects have an inherent resonance and that this was somehow replicated in humans' first words. The "bow-wow" or "cuckoo" theory, attributed to the eighteenth-century German philosopher Johann Gottfried Herder, considers the first words to be emulations of animal calls and bird cries. Herder assumed the sound of each creature evolved into a signal to distinguish it, but he did not clarify how words for inanimate or soundless objects came into being. Most words today have an arbitrary form, fixed by convention, so the theory cannot be regarded as a complete explanation for the origin of language.

Nevertheless, present-day researchers Pamela Perniss and Gabriella Vigliocco have expanded on the notion of mimicking the environment and suggested it may well be that onomatopoeia was the first step in mapping linguistic labels to concepts. By using "imitative representations of real objects and actions", the first words would have been more easily linked to their meaning in the memory. Perniss and Vigliocco have also proposed that iconicity (a similarity between linguistic form and meaning) may have played a pivotal role in establishing **displacement**, i.e. the ability of language to refer to something that is not immediately present. For example, to communicate the intention to go hunting, humans might have produced vocal imitations of previous hunting expeditions to convey the intention [9].

> **Displacement**: Feature of language that allows us to communicate ideas that are remote in time and space.

Thus, iconicity could have served as the bridge that allowed the shift from talking about something present in the world to talking about something memorised in the mind. According to Perniss and Vigliocco, one of the driving forces behind this shift would have been group size. Advanced social and cultural development in humans would have led to larger group sizes and increased cooperative interaction and information sharing, which would have required talking about matters not immediately present [9]. Others have attributed displacement to advanced human social cognition. British clinical psychologist Simon Baron-Cohen and colleagues [10] argue that **theory of mind** (the ability to judge other people's mental states) was a prerequisite for language. They base this on evidence from children's language development, which depends on understanding others' intentions to infer the meaning that speakers wish to convey. Moreover, there is evidence that connected brain areas control language and theory of mind. For example, the temporoparietal junction (TPJ) not only assists in the perception and reproduction of words and the learning of new vocabulary, but also contains areas that specialise in recognising faces and voices as well as theory of mind. Since all of these areas are in close proximity to each other, it seems likely that they interact with one another. We will revisit the concept of theory of mind in Chapter 9.

Theory of mind: Ability to judge other people's mental states, intentions and communicative goals. Implicated in pragmatic phenomena, such as indirect replies, irony and humour.

3.2 The Directive Function

Next on Leech's list is the "directive" function of language, which refers to utterances used to influence others. Today, these can be seen in the form of commands, requests or suggestions. They could have started as words (or sounds) to signify ideas such as *run!* or *help!* during hunting. Eventually, two words would have been combined to give more precise instructions (*Follow bear!*). This is in line with the "synthetic model" of language evolution, which assumes evolving language started with single items and formed structures by combining them. The first Greek grammars point to the primary role of nouns and verbs. Aristotle divided the clause into two parts: the subject, which identifies "who" or "what" is performing an action, and the predicate, which describes what action the subject is performing. As we shall see later in the book, the supremacy of nouns and verbs is reflected in children's first two-word utterances, e.g. *Mummy eat*, which typically consist of a noun and a verb. Moreover, research now suggests that our cousins, the apes, possess a form of "mental grammar" and tend to visualise the world in terms of subjects and predicates, indicating a cognitive framework similar to that of humans.

The need to give commands and coordinate action could well have been the original trigger to distinguish between nouns and verbs and join them together in a consistent way. Some theories of language evolution propose that language emerged as an adaptation to facilitate cooperation among individuals within a community or group. In his book *The Language Instinct* (1994), Pinker describes language as a human capacity created by evolution to address the specific challenge of communication within hunter-gatherer groups. The ability to give and follow orders would have been crucial for coordinated group activities such as hunting, gathering or defence. Being able to give clear instructions about hunting strategies or coordinating activities during a shared task would enhance the survival chances of the group.

In the social context, the directive function is often associated with authority and hierarchy and may have played a role in the development of hierarchical social structures. The directive function could also be seen as an adaptive response to the need for effective problem-solving within social groups. Giving orders would have enabled individuals to communicate solutions, strategies and instructions to address challenges collectively, contributing to the survival and success of the community. Thus, overall, the ability to give and understand directives could have played a significant role in the coordination, cooperation and survival of early human communities.

3.3   The Social Function

As already indicated at the beginning of the chapter, perhaps the most important task of language is to build and maintain social contact. Present-day examples of the "social" function of language include expressions such as *How are you?* or comments about the weather when being used to start up interaction with someone. These phrases could be seen as fulfilling the same function as animal contact calls, which are used to keep groups united (and help reconnect visually separated individuals). Robin Dunbar's "vocal grooming hypothesis" [11] argues that strong social ties bring fitness advantages to individuals, but that grooming—the primary mode of socialisation in non-human primates—demands a considerable amount of time and energy. We, therefore, evolved to replace this physical grooming with low-cost vocal sounds, or "vocal grooming", which kept hands free for other tasks. Dunbar suggests vocal grooming then evolved into vocal language, although the theory does nothing to explain the transition from holistic "sing song" to syntactic speech.

Another theory based on social interaction is Dean Falk's "Putting-Down-The-Baby" theory [12]. It states that with the trend towards larger brains in early Homo and hence earlier births, human mothers were not able to move around and forage for food with their babies on their backs since human babies were too undeveloped to cling on. Therefore, mothers frequently left their babies on the ground and, in order to reassure them that they were not being abandoned, started using contact calls. These vocal exchanges between early human mothers and infants set in motion a sequence of events that eventually led to humans' first words.

As a development of the "Putting-Down-The-Baby" theory, Oren Poliva's neurally-based "From-Where-To-What" model [13] [14] specifies how human speech could have developed from contact calls into the fully fledged language via a transitory "prosodic" phase (where meaning is conveyed holistically by stress and intonation rather than by meaningful segments). This is a variation of the "analytic model" discussed earlier in the chapter. Here is a brief outline of the model (Fig. 3.2).

The name of Poliva's From-Where-To-What model stems from the two auditory streams found in the brains of humans and other primates. The auditory ventral stream (AVS) contains various circuits specialised for perceiving and recognising sounds and is thus called the auditory "what" stream. Sound recognition is highly developed in many mammals, being used to identify prey, predators or potential mates. In

1. Speech evolved in early hominins for the purpose of exchanging contact calls between mothers and their offspring to find one another in the event they became separated: Mother: *koo*   Child: *koo*
2. The contact calls were modified with intonations to express either a higher level of distress to signal trouble (*ko!*) or a lower level of distress to signal satisfaction (*koo*).
3. The use of two types of contact calls enabled the first question-answer conversation. For example, when an infant expressed a low-level distress call prior to eating berries, his/her mother could have responded with a high-level distress call, which indicated the food is dangerous, or a low-level distress call, which indicated the food is safe. Eventually, the infant emitted the question call and waited for an appropriate answer from its mother before proceeding with an intended action. This conversation structure could be the precursor to present-day **yes/no questions**: Child: *koo?* (= Is it ok?) Mother: *ko!* (= No, it's dangerous) Child: *koo?* (= Is it ok?) Mother: *koo* (= Yes, it's safe)
4. Likewise, a call using intonations to express a high level of distress could have served   as a precursor to contemporary **commands**, while a call using intonations to express a low-level of distress could have served as the precursor to present-day **wh- questions**: *ma!* (= Mummy, come here now!) *maa?* (= Mummy, where are you?) This transition could be compared to the ability of present-day infants who, at the single-word stage of language development, can already signal a command (*Mummy!*) or a question (*Mummy?*) through intonation.
5. Over time, the improved use of intonations and vocal control led to the invention of unique calls (proto-words) associated with distinct objects. Thus, *koo* might have produced *kroo*, *kraa* and *kree*. Children first learned words from their parents by imitating their lip movements, and then words were eventually recognised and memorised for retrieval. As the need for more words increased, syllables were combined to form **polysyllabic words**. Finally, sequences of syllables provided the framework for communicating with **sequences of words** (i.e., sentences).

Fig. 3.2 Poliva's model of language evolution: from contact calls to language (based on Poliva 2015, 2016)

primates, the auditory dorsal stream (ADS) is developed for locating sounds and is thus referred to as the auditory "where" stream. In humans, the ADS is additionally responsible for processes associated with vocal production. According to Poliva [15], structural changes to the ADS equipped early Homo with the superior vocal control necessary for speech.

All these theories of language emerging from social interaction assume that language started out in the vocal form. However, when we consider that gestures (manual and facial) are an inherent and important part of primate communication, and that very young children use pointing and symbolic hand gestures before they can speak, it seems logical that speech might have developed from gesture. Corballis' "Gesture-first hypothesis" of language origins [16] argues exactly this, that humans were first able to communicate symbolically by gesture, before vocal elements were added. Evidence from neuroscience has shown that gestures and speech are processed using similar neural pathways. It has been hypothesised that those parts of the brain that initially supported the pairing of gesture and meaning were adapted in human evolution for the pairing of sound and meaning in speech. The result is that speech and gesture are inherently linked in the brain and function in an efficiently wired and choreographed system [17]. This instinctive linking of speech with gestures is further revealed by the fact that blind individuals use gestures during conversation despite lacking visual input for such behaviour.

Others believe that gesture and speech evolved together as manifestations of a single **multimodal** cognitive system, with neither **modality** taking precedence over the other. Rather, human language is the outcome of a long process of modification to a complex multi-component communication system. Gesturing is an integral part of utterances in that it provides additional nuances of meaning. According to gesture expert Adam Kendon, manual gesturing can convey further details about the dimension or position of objects we are talking about, or the manner of an action, thus giving visible form to abstract concepts and relationships [18]. Facial gesturing can also signal dimension and space, as well as reveal the attitude of the speaker.

Multimodal: Involving a variety of modes (e.g. vocal, auditory, gestural, visual).

Modality: Channels involved in different forms of language. Speech involves the vocal-auditory modality while gesturing/sign language involves the gestural-visual modality.

The question arises what caused the vocal-auditory modality to develop further in humans. Comparative studies of different species of monkeys have shown correlations between the complexity of vocal repertoires and the complexity of social organisation. The "social complexity hypothesis" [19] proposes that societies that are larger and have differentiated social roles and more complex interaction networks, including maintained pair relationships, will have a wider and more diverse repertoire of communicative signals, both vocal and visible. If this is so, this means that any evolutionary account of human language should take account of the evolution of social complexity in humans. The role society plays in shaping language will be investigated further in Chapter 5.

3.4 The Informational Function

Leech's "informational" function of language refers to statements of fact that are intended to provide readers or listeners with useful information. In social group-living animals, there are obviously situations where information sharing is vital for group survival, such as when it is necessary to alert others of danger. The "uh-oh" or "warning hypothesis" suggests language could have evolved from warning signals, as used by many animals to alert other group members of some approaching threat. Humans would have been able to build on this when they began to differentiate calls. For example, one sort of cry could signal that lions had been spotted, and another one could indicate a snake. Eventually, humans would have somehow realised that the different sounds they were using could be used to name things out of sight. This **"naming insight"** would have been a huge leap in the evolution of language and no doubt led to an explosive growth of vocabulary, which would have required enhanced vocal control to produce more and more sounds and enhanced memory to memorise all the new words.

> **Naming insight**: Pivotal moment in the evolution of language when humans realised objects in their environment could be identified with labels, laying the foundation for the development of vocabulary. Some believe the naming insight was the crucial moment that distinguished human language from other animal communication systems.

Nevertheless, some believe it is highly unlikely that language developed for the sole purpose of transferring knowledge since communication relies heavily on inferential processes and serves goals other than the faithful transmission of **propositional** information. In a ground-breaking article on animal communication, British evolutionary biologist Richard Dawkins and zoologist John Krebs [20] suggested that the most significant mental ability for the development of human language was the ability to tactically deceive others. Deception involves bending the truth and is often done for personal gain. In the natural world, animals may deceive predators by mimicking a more dangerous animal. It has been shown that some monkeys can use fairly sophisticated forms of deception in order to obtain food, but only we were successful in exploiting this ability linguistically. The "deception hypothesis" of language origins, as proposed by cognitive scientist Thom Scott-Phillips [21], argues that humans first learned to suppress involuntary cries and feign their true mental state. In turn, human language exploited the ability to displace emotions and used it to communicate ideas that are remote in time and space (displacement).

> **Propositional meaning**: Part of meaning conveyed by a sentence that relates to some state of affairs in the world. Propositional meaning remains the same regardless of when or where the sentence is uttered.

3.5   The Aesthetic Function

Leech's last function of language is the "aesthetic" function, which he applied to words and sentences used in poetry and song for their artistic and musical value. The "yo-he-ho hypothesis" proposes that language arose out of the rhythmic chants uttered by people engaged in communal labour. While this notion may seem a little far-fetched (it might explain some of the rhythmic features of language, but it does not really account for where words came from), musicality does seem to be deeply ingrained in us. Some researchers believe musicality may have evolved as an adaptive trait, serving various functions such as the expression of emotions, social bonding and communication. Some say it was essential for language learning. As we shall see later in the book, infants' language acquisition begins with discriminating the sounds and rhythm of language. The fact that prosody is handled primarily by the right hemisphere of the brain while mainstream language is processed in the left hemisphere seems to imply that musicality was a pre-existing component that has been integrated into speech.

In the previous chapter, we saw that language can be broken down into segments, including words and speech sounds. We also saw that spanning across these segments are suprasegmental features, such as stress, rhythm and intonation. These not only provide a pulse to speech but also a wealth of information over and above the literal meaning of the words uttered. Let us look at these features individually. **Intonation** (falling and rising pitch, otherwise known as speech melody) contributes substantially to meaning. Firstly, it can indicate the grammatical function of an utterance (such as whether it is a statement or question). Compare the intonation pattern of the word *here* in the following:

Tom's already ↘ here!   (falling pitch for a statement)

Tom's already ↗ here?   (rising pitch for a question)

The pitch pattern used in a question tag can even signal the degree of certainty the speaker feels. Compare the following:

They missed the train, ↘ didn't they? (falling pitch for certainty)

They missed the train, ↗ didn't they? (rising pitch for uncertainty)

In addition, intonation can reveal attitudes and emotions. A high fall in a statement is the unmarked (default) pattern and neutral in tone, whereas a low fall is cooler and more distant, sometimes even sarcastic. Compare:

I'm delighted to ↘ meet you

I'm delighted to ↘ meet you

> **Intonation**: Pitch contour of speech that can indicate the grammatical function of an utterance (e.g. statement vs. question), and also the attitudes and emotions of the speaker.

Many languages can additionally use different pitch patterns, or tones, to create different words. For example, in the Ewe language of Ghana, *tó* (pronounced with a high tone) means "ear" and *tò* (pronounced with a low tone) means "buffalo"! Mandarin Chinese uses four tones—level, rising, fall-rise and falling—to distinguish words. For example, *mā* means "mother", *má* "hemp", *mǎ* "horse" and *mà* "scold".

Stress (emphasis placed on certain words and syllables) is achieved through changes in loudness, pitch and vowel length, and serves several important purposes. For example, stress can be placed on particular words to highlight new information:

Tom's here! (Now the party can start!)

or else to signal a contrast with previous information:

Tom's here! (I thought he was on holiday!)

> **Stress**: Emphasis placed on a syllable in a word (word stress) or at different points in a sentence (sentence stress) through increased loudness and vowel length and a rise in pitch. Word stress is usually fixed, whereas sentence stress can be varied to highlight salient information or create contrast.

Rhythm (the pattern of timing formed by placing stress on certain syllables) is an essential component of human speech; it not only helps us understand the speech of others, but also assists children in acquiring language. This is due to the fact that in connected speech, content words (nouns, verbs, adjectives and adverbs), which carry a high information load, receive more *stress* than function words (such as articles, prepositions, pronouns and conjunctions), which convey relatively little information. This allows us to pay attention to the most important information for further processing. In the following sentence, notice how unstressed "to" (a preposition) is reduced from /tu:/ to /tə/ or simply /t/:

I've heard that Tom and Sue are mo-ving to France.

In some languages (e.g. English, Dutch, German, Danish, Russian), the stressed syllables tend to occur at roughly equal intervals of time—try saying the above sentence out loud while clapping on the stressed syllables. Your claps should be nice and regular. If there are more unstressed syllables in between, these will be compressed, and if there are less, they will be expanded. Try saying the following two sentences, again clapping on the stressed syllables, using the same amount of time for each sentence:

Tim's	bought	a _flat_	in _town_
Ti-mo-thy's	_pur_-chased	an a-_part_-ment	in the _ci_-ty

Other languages (e.g. French, Spanish, Greek, Polish, Hindi) work on a different principle, where each syllable is approximately the same length regardless of stress. This shows that musicality can develop in different directions and is dependent, as is vocabulary and grammar, on conventions.

Rhythm: Pattern of timing in an utterance caused by the succession of stressed and unstressed syllables. Some languages have an equal time lag between stressed syllables while others put equal timing on each syllable.

While some see prosody as a by-product of language, others assume language was originally more musical than linguistic. As mentioned earlier, Darwin's insightful "musical origins" hypothesis has been making a comeback and continues to inspire work in a number of areas, whether it be comparing the neurological foundations of human language and birdsong or making computer simulations of holistic/analytic transitions.

Others have added more functions of language to Leech's list, including the educational (a means to understanding and exploring the environment), the imaginative (for the purpose of telling stories and creating fiction) and as an instrument of thought and reasoning—why else do we speak to ourselves in a private monologue, if not to regulate internal thought? According to the Russian and Soviet psychologist Lev Vygotsky (1896–1934), just as we construct physical tools to manage the physical environment, language is a culturally constructed artefact to manage our mental environment. In other words, language is a tool for negotiating meaning. For such higher level functions of language, which go beyond communication for immediate needs, language relies heavily on its symbolic quality. **Symbolism** not only enables us to represent abstract concepts, ideas or complex thoughts, but also allows us to refer to things out of sight, the past or future, the possible or hypothetical. For many, this is the decisive leap that distinguishes human language from other animal communication systems.

Symbolism: Practice of representing objects and ideas in symbolic form. Symbolism allows us to refer to things out of sight, the past or future, the possible or hypothetical.

3.6 Chapter Summary

By looking at the multiple functions of language today, linguists have proposed a number of hypotheses to explain the emergence of language. Some, e.g. Dunbar [11], defend a "speech-first" view, while others, e.g. Corballis [16], prefer the "gesture-first" hypothesis. Some assume language evolved out of the need to cooperate, others out of the need to deceive. Some say musicality was critical to the development of language, others see it as a by-product. It seems unlikely that any one hypothesis can describe the whole process; it is more likely that multiple mechanisms, working together or one after another, contributed to the development of language. Though it seems very plausible that language evolved for facilitating social bonding in large and complex social groups, this must have been accompanied by an ability to think beyond the present and understand the intentions of others. Thus, language is multimodal and multifaceted, involving gesture, musicality and a high degree of cognitive and social intelligence.

In the next chapter, we will look at the environmental conditions and biological adaptations that were necessary for human language to develop. We will start with the physical environment at the time when humans were still biologically closer to apes and roaming the forests of Africa.

References

1. Wray, A. (1998). Protolanguage as a holistic system for social interaction. *Language and Communication, 18* (1), 47–67. https://doi.org/10.1016/S0271-5309(97)00033-5
2. Arbib, M. A. (2005). From monkey-like action recognition to human language: An evolutionary framework for neurolinguistics. *Behavioral and Brain Sciences, 28*, 105–167 https://doi.org/10.1017/s0140525x05000038
3. Darwin, C. (1871). *The descent of man.* John Murray.
4. Jespersen, O. (1922). Language, Its Nature, Development, and Origin. *The American Journal of Philology, 43*(4), 370–373.
5. Fitch, W. T. (2013). *Musical protolanguage: Darwin's theory of language evolution revisited.* In J. Bolhuis & M. Everaert (Eds.), *Birdsong, speech, and language: Exploring the evolution of mind and brain* (Ch. 24, pp. 489–503). MIT Press.
6. Bickerton, D. (1990). *Language and Species.* University of Chicago Press
7. Jackendoff, R. (2002). *Foundations of language: Brain, meaning, grammar, evolution.* Oxford University Press
8. Tallerman, M. (2007). Did our ancestors speak a holistic protolanguage? *Lingua* 117, 579–604 https://doi.org/10.1016/j.lingua.2005.05.004
9. Perniss, P., & Vigliocco, G. (2014). The bridge of iconicity: From a world of experience to the experience of language. *Philosophical Transactions of the Royal Society B Biological Sciences, 369*, 20130300. https://doi.org/10.1098/rstb.2013.0300
10. Baron-Cohen, S., Tager-Flusberg, H., & Cohen, D. J. (2000). *Understanding Other Minds: Perspectives from Developmental Cognitive Neuroscience (2nd ed.).* Oxford University Press
11. Dunbar, R. (1996). *Grooming, gossip and the evolution of language.* London: Faber and Faber
12. Falk, D. (2004). Prelinguistic evolution in early hominins: whence motherese? *Behavioral and Brain Sciences, 27*(4), 491–503 https://doi.org/10.1017/s0140525x04000111

13. Poliva, O. (2015). From Where to What: A Neuroanatomically Based Evolutionary Model of the Emergence of Speech in Humans. *F1000Research* 4:67 https://doi.org/10.12688/f1000rese arch.6175.1
14. Poliva, O. (2016). From Mimicry to Language: A Neuroanatomically Based Evolutionary Model of the Emergence of Vocal Language. *Frontiers in Neuroscience*, 10, Art 307 https://doi.org/10.3389/fnins.2016.00307
15. Poliva, O. (2017). From where to what: A neuroanatomically based evolutionary model of the emergence of speech in humans. *F1000Research* 6:67. https://doi.org/10.12688/f1000rese arch.6175.3
16. Corballis, M. C. (2002). *From hand to mouth*. Princeton University Press
17. Wikipedia .(2023). Gesture/Neurology. https://en.wikipedia.org/wiki/Gesture#Neurology
18. Kendon, A. (2016). Reflections on the "gesture-first" hypothesis of language origins. *Psychon Bull Rev 24*, 163–170. https://doi.org/10.3758/s13423-016-1117-3
19. Freeberg, T. M., Dunbar, R. I. M., & Ord, T. J. (2012). Social complexity as a proximate and ultimate factor in communicative complexity. *Phil. Trans. R. Soc. B, 367*, 1785–1801 https://doi.org/10.1098/rstb.2011.0213
20. Dawkins, R., & Krebs, J. R. (1978). Animal signals: information or manipulation? In J. R. Krebs and N. B. Davies (Eds.) *Behavioural Ecology* (Oxford: Blackwell Scientific Publications), 282–309
21. Scott-Phillips, T. C. (2006). Why talk? Speaking as selfish behaviour. In A. Cangelosi, et al. (Eds.). *The Evolution of Language: Proceedings of the 6th International Conference on the Evolution of Language* (pp.299–306). Singapore: World Scientific.

Chapter 4
How the Physical Environment Shaped Language

In this part of the book, we will look at the various environmental conditions that led to language. The chapter starts off on the geological timescale and investigates the changes in the physical environment that led to anatomical, physiological and neural adaptations in Homo. These, in turn, led to an immense increase in cognitive capacities that favoured the development of language. Chapters 5 and 6 will deal with the equally significant social and cultural conditions that contributed to language.

4.1 Tectonics and Climate

In the **Miocene** (extending roughly from 23 to 5 million years ago), there was a great diversity of ape species, with dozens of fossilised species found across Africa, Europe and Asia. The transition from an ape-like to a more human-like anatomy is very evident in fossils from the late Miocene, most of which have been found in Africa's **Great Rift Valley**. It has been suggested that this area, born out of exceptional tectonic uplift and subsequent down warp, creating rain-starved basins, experienced accelerated aridification and was therefore a major driver of hominin evolution and **speciation**. The well-known "savannah hypothesis" proposes that the new savannah environment was unsuitable for tree monkeys and favoured those species that could stand on their back limbs [1]. Around 3.5 million years ago, there were several species of Australopithecus that had developed this ability, possibly through the habit of standing on tree branches to reach otherwise inaccessible fruit. The "aridity hypothesis" [2] expands on the savannah hypothesis by suggesting that periods of accelerated aridification caused thresholds in evolution and major hominin speciation events.

© The Author(s), under exclusive license to Springer Nature Switzerland AG 2024   49
J. Dornbierer-Stuart, *The Origins of Language*,
https://doi.org/10.1007/978-3-031-54938-0_4

Miocene: Geological epoch extending from about 23 to 5 million years ago. The late Miocene saw the emergence of Hominina (hominins).

Great Rift Valley: Term often used for the East African Rift System that developed around the onset of the Miocene. The shift to more arid, open conditions led to major steps in the evolution of hominins.

Speciation: Evolutionary process by which populations evolve to become distinct species. May occur through splitting of lineages (e.g. by migration) or through genetic mutation and recombination within a population.

It is thought that shorter episodes of extreme climatic variations in East Africa particularly favoured rapid brain expansion. These fluctuations in climate would have resulted in changes in vegetation, water availability and the distribution of resources. The "variability selection hypothesis" [3] proposes that these periods of rapid and unpredictable environmental changes placed unique and demanding selective pressures for behavioural flexibility on early human populations. Over time, these selective pressures would have driven the expansion of brain size and the development of complex cognitive and behavioural traits (such as language) that are characteristic of modern humans.

4.1.1 Bipedalism

As the ancestors of early humans descended from the trees, sometime between 4.2 and 3.5 million years ago, during the **Pliocene**, they had to adapt to survive. Herbivorous hominins took to hunting and eating meat, which meant competing with the emerging big cats. In his engaging book *Sapiens: A Brief History of Humankind* (2014), Israeli historian Yuval Noah Harari claims that the upright stance and **bipedalism** made it easier to scour the savannah for prey and predators, and left arms free for other pursuits, such as stone throwing or signalling. And the more hands were used, the more evolution brought about an ever greater concentration of nerves and fine-tuned muscles in the palms and fingers [4]. As a result, humans were able to use and produce sophisticated tools by around 2.5 million years ago, at around the start of the **Pleistocene** (Fig. 4.1).

Fig. 4.1 Australopithecus family from the Early Pleistocene (by Mauricio Antón)

Pliocene: Geological epoch extending from around 5 to 2.5 million years ago. The early Pliocene saw the emergence of Australopithecus, from which Homo emerged.

Bipedalism: Ability to walk on two feet. Humans are habitual bipeds whose normal method of locomotion is two-legged.

Pleistocene: Geological epoch, commonly known as the Ice Age, extending from around 2.5 million to 12,000 years ago. The early Pleistocene saw the emergence of Homo.

While the direct connection between bipedalism and language evolution is complex and not fully understood, there are several ways in which the development of bipedalism might have facilitated the emergence and evolution of language in early humans. Firstly, the resulting increase in dexterity and manipulation abilities could have laid the foundation for the development of gestures, which could have served as precursors to language (see Corballis' "gesture-first hypothesis" in Sect. 3.3). Early forms of communication might have relied on manual gestures that were used to convey simple messages and intentions. Eventually, gestures and vocalisations would have been integrated into a more complex linguistic system. Secondly, an increased use of tools would have required greater spatial awareness and cognitive skills. These cognitive abilities could have been co-opted for spatial reasoning and abstract thinking, both of which are important for language. Finally, the upright posture would have facilitated more direct and sustained face-to-face interactions among individuals. The resulting enhanced social interaction could have provided a platform for the development of more sophisticated forms of communication, including the expression of emotions, intentions and social bonds, all key components of language.

4.2   Reconfiguration of the Vocal Tract

As well as freeing the hands, the upright stance would have altered the shape of the mouth and **vocal tract**. According to researchers Ghazanfar and Rendall [5], one of the most obvious differences in the vocal anatomy of humans and non-human primates is the descended position of the **larynx** in the human vocal tract. The outcome is a double-tubed tract consisting of the oral cavity, which is present in all primates, and an enlarged pharyngeal cavity found only in humans (see Fig. 4.2). This double-tube configuration, combined with a flexible curved tongue and the ability to make rapid jaw and lip movements, affords humans remarkable articulatory agility when vocalising.

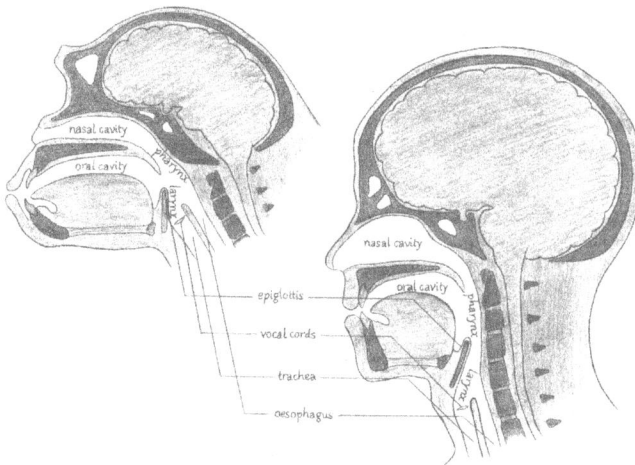

Fig. 4.2 Vocal tract of the chimp and modern human (by William Duffy)

> **Vocal tract**: Passageway used in the production of speech, above the larynx and including the oral, nasal and pharyngeal cavities.

> **Larynx**: Voicebox housing the vocal cords, which manipulate the pitch and volume of the voice.

Ghazanfar and Rendall note that freedom of articulatory movement leads to dynamically changing resonances in the vocal tract, and it is precisely this property that is required to define the phonemes of contemporary languages. This is why the descent of the larynx in humans has long been regarded as a crucial anatomical adaptation for language and was even thought by some to mark the origin of language in early humans [5]. However, while the descent of the larynx certainly had an impact on the development of language, most today would agree it is unlikely to be the sole or even primary factor. Above all, such a complex behaviour as language would have required substantial development of the brain.

4.3 More Time to Develop

There was a major drawback in adjusting to the upright position, although language ultimately benefitted from the situation. According to Harari, "An upright gait required narrower hips, constricting the birth canal – and this just when babies' heads were getting bigger and bigger. Women who gave birth earlier … fared better and lived to have more children. Natural selection consequently favoured earlier births" [4]. As a result, humans are born with many vital systems under-developed, and so remain dependent on their elders for many years. Consequently, evolution also favoured those able to form strong social bonds. In addition, being born "prematurely" means brains are highly plastic and remain sensitive to environmental stimuli for longer, which is crucial for **neurodevelopment**. The upshot is that humans can be nurtured and socialised to a much greater degree than any other animal. On the downside, we take much longer to reach full maturity because of our larger brains that take longer to develop.

> **Neurodevelopment**: Formation of the neurological connections and pathways in the brain that allow a highly complex set of behaviours, including language.

This new social environment in which our early human ancestors lived would have placed increasing demands on cognitive abilities, and larger brains may have provided the processing power needed to meet these challenges. As a result, archaic humans had much larger brains compared to other animals. Present-day mammals weighing 60 kg have an average brain size of 200 cm³. Fossil evidence has revealed that the earliest humans, some 2.5 million years ago, had brains of about 600 cm³. In some ways, this presented a disadvantage for survival: larger brains required disproportionately much more fuel than smaller brains and so more time was needed for hunting. In addition, human brains grew at the expense of muscles, making it more difficult for humans to defend themselves [4]. For the next two million years, human brains kept growing and growing, no doubt assisted by the development of salivary amylase (allowing for the more efficient digestion of starch-rich foods such as roots and tubers when meat was scarce) as well as the great invention of fire, some 1.5 million years ago, which enabled humans to obtain calories more efficiently. Some believe there is a direct link between the invention of cooking and the expansion of the human brain [4]. The modern human brain today has reached a volume averaging around 1200 cm³.

4.4 Restructuring of the Brain

According to Dr. Marcia Ponce de León from the Institute of Anthropology at the University of Zurich, computed tomography analyses of a range of fossil skulls suggest that modern human brain structures started to emerge approximately 1.5 million years ago, around the time of Homo erectus. During this time, culture in Africa became more sophisticated and diverse, as reflected by the discovery of a variety of stone tools. Ponce thinks it is quite possible that the earliest forms of human language developed during this period as well. She explains that, apart from its size, the modern human brain distinguishes itself from that of great apes primarily through the location and organisation of certain brain regions, especially those in the frontal area, which are responsible for the planning and execution of complex patterns of thought and action and, ultimately, also for language [6].

Traditionally, vocal control in humans has been linked to **Broca's area**, a cluster of interconnected areas located in the inferior frontal gyrus of the frontal lobe of the dominant hemisphere of the brain (usually the left side). Together, these areas not only control the muscles associated with articulation but also enable us to order concepts and process syntax. The area was discovered when the neurologist Paul Broca reported the loss of speech in patients after injury to this part of the brain. Speech comprehension has been linked to **Wernicke's area**, located in the superior temporal gyrus of the temporal lobe of the dominant hemisphere of the brain. In functional brain imaging experiments, this site has been associated most frequently with auditory word recognition.

Broca's area: Cluster of interconnected areas located in the inferior frontal gyrus of the frontal lobe of the brain. Associated with vocal control and syntax.

Wernicke's area: Cluster of interconnected areas located in the superior temporal gyrus of the temporal lobe of the brain. Associated with speech comprehension.

According to neuroscientists Tremblay and Dick [7], while the above model is still relevant to clinicians, it perpetuates the belief that there exists neural tissue specifically dedicated to the task of **language processing**. They take an alternative view, "that language is, at least in part, an overlaid functional system that gets what service it can out of nervous tissues that have come into being and are maintained for very different ends than its own" [7]. In other words, language depends heavily on domain-general neural resources that just happened to be recruited for language.

Language processing: Mental processes involved in understanding, producing and acquiring any form of language, including spoken, written and sign language.

Contemporary models of the neuroanatomy of speech include areas such as the auditory cortex, responsible for processing auditory information, and the motor cortex, which is involved in controlling the muscles involved in speech production. There is also evidence from neuroimaging that speech comprehension tasks activate the visual cortex, particularly in blind people, but also in sighted individuals. This seems to be clear evidence of multisensory connectivity in the brain. A final area worth mentioning is the inferior parietal lobule (IPL), also known as "Geschwind's territory", which is involved in the interpretation of sensory information. At the intersection of the hearing, touch and vision parts of the brain, it is thought to play a key role in identifying multisensory abstract qualities of objects or ideas and associating words with them [8]. This seems to be supported by the fact that across many languages certain sounds are used to mimic certain abstract qualities. For example, bilabial consonants (e.g. /m/ and /b/) are often linked to words for "soft" and "smooth" while the trilled r is frequent in words for "rough". While the precise link between these linguistic phenomena and the inferior parietal lobule is not well-established, it is clear that language processing involves more than Broca and Wernicke and is part of a broader neural and functional sensorimotor network.

Figure 4.3 shows a more comprehensive architecture for language, encompassing a number or regions not considered by the classic model. These regions, all working in tandem, would have provided Homo sapiens with a significant cognitive advantage in expressing ideas, coordinating actions and transmitting knowledge to others.

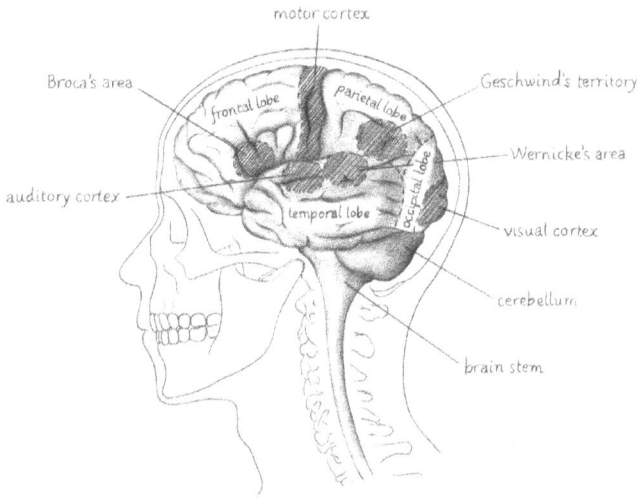

Fig. 4.3 Speech areas of the brain (by William Duffy)

4.5 An Explosion of Culture

According to Harari, several species of humans were hunting large game on a regular basis 400,000 years ago, but it was only with the rise of Homo sapiens, roughly 100,000 years ago, that humans jumped to the top of the food chain [4]. One reason this happened so suddenly could well have been sophisticated language. Some animal communication systems can involve a number of signals, including vocalisations and gestures, but they are simply less complex and therefore less productive. The fact that we can combine a limited number of sounds and signs to create an infinite number of sentences means that we can "ingest, store and communicate a prodigious amount of information about the surrounding world" [4]. This wealth of information would have allowed **culture** to skyrocket, and, in turn, an explosion of culture would have meant that language could develop even further.

> **Culture**: Set of knowledge, customs and habits acquired socially, through imitation and learning. Includes "virtual realities" such as concepts, beliefs, institutions and language.

Figure 4.4 provides a summary and rough timeline for some of the most significant adaptations in early humans that allowed language to emerge and proliferate:

It could be argued that until the explosion of culture, roughly 50,000 years ago, it was primarily nature that provided the biological framework for language; it was our

Formation of Great Rift Valley in East Africa	Early humans adapt to savannah: upright stance, hunting and meat-eating, reconfiguration of vocal tract	Humans use sophisticated tools, brains enlarge, evolution favours earlier births	Modern human brain structures emerge, humans start to exploit fire	Large game hunting, anatomically modern humans	Homo sapiens top of food chain, symbolic language which is highly productive	Explosion of culture
8 million years ago	4 million years ago	2.5 million years ago	1.5 million years ago	300,000 years ago	100,000 years ago	70,000 years ago

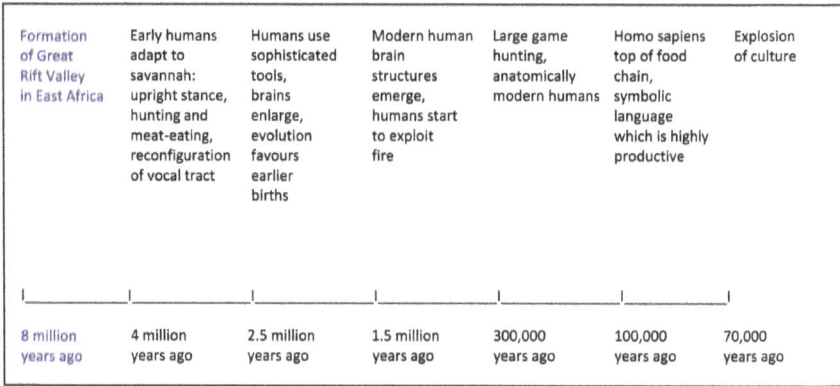

Fig. 4.4 Adaptations in early humans leading to language

physical features and mental abilities that primarily paved the way for speech. As we shall see in Chapter 9, it is now largely recognised that the special cognitive abilities necessary for language have their roots in the animal world. Since humans and chimpanzees have a common ancestry and, at some point, diverged into two distinct species from a **last common ancestor (LCA)**, scientists can infer, by comparing the brain structure, vocal apparatus and genetic makeup of humans and chimps, the evolutionary changes that may have influenced the development of language. Today much progress is being made in ascertaining evolutionary **continuity** between humans and their animal predecessors. On the one hand, language relies on relatively simple ancient mechanisms such as sensory-motor coordination and imitation found in many animals. On the other hand, there is growing evidence that the building blocks of more sophisticated mechanisms such as **combinatorial phonology** (combining acoustic signals in meaningful ways), can also be found in the animal world.

Last common ancestor (LCA): In biology, the most recent individual from which two species (e.g. chimps and humans) are descended.

Continuity: View that language has evolved from precursors in the animal world.

Combinatorial phonology: Combining meaningless elements (speech sounds) to form meaningful elements (morphemes and words).

However, this is only half the story. As language spreads across a community and is transmitted from one generation to the next, it gets reshaped by its users and learners. As researchers Zuidema and de Boer point out, even if we assume the building blocks of combinatorial phonology are inherited from other species, it only finally emerges when there is repeated interaction between individuals and a growing number of concepts that need to be expressed [9]. In other words, combinatorial phonology is as much a social and cultural phenomenon as it is the product of genes. The significance of social interaction and cultural transmission for theories of language evolution will be explored further in the next two chapters.

4.6 Chapter Summary

In this chapter, we began by discussing how the physical environment shaped the human anatomy, creating the pre-conditions for the cognitive developments that made our language possible. Herbivorous hominins took to hunting and eating meat and standing on two feet to better scan the land for game and enemies. As well as freeing the hands, the upright posture altered the shape of the mouth and throat, allowing humans considerable articulatory agility when vocalising. It also narrowed the hips, meaning children had to be born earlier—before heads and brains were fully developed. This caused children to be more dependent on elders, which provided the perfect conditions for language learning. Through social interaction and cultural transmission, language would have developed further.

In the next chapter, we shall take a closer look at the social and cultural conditions that paved the way for language.

References

1. Maslin, M. A., Brierley C. M., Milner, A. M., Schultz, S., Trauth, M. H., & Wilson, K. E. (2014). East African climate pulses and early human evolution. *Quaternary Science Reviews, 101*, 1–17. https://doi.org/10.1016/j.quascirev.2014.06.012
2. de Menocal (1995), as cited in Maslin et al. (2014). East African climate pulses and early human evolution. *Quaternary Science Reviews, 101*, 1–17. https://doi.org/10.1016/j.quascirev. 2014.06.012
3. Potts (1998), as cited in Maslin et al. (2014) East African climate pulses and early human evolution. *Quaternary Science Reviews, 101*, 1–17. https://doi.org/10.1016/j.quascirev.2014. 06.012
4. Harari, Y. N. (2014). *Sapiens: A brief history of humankind*. Harvill Secker.
5. Ghazanfar, A., & Rendall, D. (2008). Evolution of human vocal production. *Current Biology, 18*(11), R457. https://doi.org/10.1016/j.cub.2008.03.030
6. Gull, T., & Nickl, R. (2021). Anthropologie: Modernes Hirn. *UZH Magazin, 2*(21), 28.
7. Tremblay, P., & Dick, A. S. (2016). Broca and Wernicke are dead, or moving past the classic model of neurobiology. *Brain and Language, 162*, 60–71. https://doi.org/10.1016/j.bandl.2016. 08.004

8. Ramachandran, V. S., & Hubbard, E. M. (2001). Synaesthesia—A window into perception, thought and language. *Journal of Consciousness Studies, 8*(12), 3–34.
9. Zuidema, W., & de Boer, B. (2018). The evolution of combinatorial structure in language. *Current Opinion in Behavioural Sciences, 21*, 138–144. https://doi.org/10.1016/j.cobeha.2018.04.011

Chapter 5
The Impact of Society and Culture on Language

At the end of the last chapter, we saw that certain features of language (such as combinatorial phonology) depended not only on biological and cognitive capacities but also on social and cultural processes. While biological accounts of language evolution focus primarily on cognitive aspects of language that are "internal" to the individual, the cultural approach sees language as an "external" object that exists in the community, and thus attempts to identify social and cultural pressures that constrain the evolution of language. It considers the role of social interaction and children's language learning and often relies on computational models to simulate the cultural evolution of language. We will begin this chapter by looking at the effect **society** and its social structures would have had on emerging language, especially regarding the need to hunt, keep peace and raise offspring. We will then explore the cultural evolution of language and the notion that language is self-evolving as it is transmitted across the community, and we will learn about some agent-based models that simulate the evolution of language. In addition, we will explore the possibility that not only genes affect language but, vice versa, language affects genes in a complex interplay between biology and culture, finishing with Kirby's loop of three adaptive systems.

Society: Community of interdependent organisms, usually of the same species. A society can offer its members benefits that would not be possible on an individual basis.

© The Author(s), under exclusive license to Springer Nature Switzerland AG 2024   61
J. Dornbierer-Stuart, *The Origins of Language*,
https://doi.org/10.1007/978-3-031-54938-0_5

5.1 The Effect of Society on Language

Because humans are social animals, there is a greater need for cooperation in order to survive. It is commonly believed that hunting large game in the savannah would have required a greater degree of planning and coordination, and this would have required more efficient communication. The development of combinatorial and hierarchical structure in language (see Sect. 2.1) would have been a major step in making communication more efficient since it ultimately allows complex ideas to be conveyed with limited resources, and those who adapted to it better would have had a better chance of survival.

In addition to a greater need to cooperate, larger groups of humans would have benefitted from forming friendships and hierarchies to keep peace and order. Gossip was one means that helped humans form larger and more stable bands, but some have argued that it was the ability to manipulate reality and create fiction, including tales of supernatural beings, that was more crucial for keeping society in check. These fictional narratives would have been possible due to the ability of language to refer to something beyond the immediate present (displacement) (see Sect. 3.1) and the use of symbolism (see Sect. 3.5). In fact, such narratives are comparable to our present-day social customs and institutions. Most of us probably have a similar regard for the constructs of marriage, the nuclear family, democracy or the Universal Declaration of Human Rights as our ancestors had for their myths and legends.

Since the advent of **symbolic culture**, humans have been able to pass on new behaviours to future generations without the need for genetic modification. According to historian Yuval Noah Harari, despite having virtually the same DNA as us, Neanderthals, with another social organisation, were unable to cooperate effectively (or develop the language skills to do so). As a result, they were unable to adapt their social behaviour to rapidly changing conditions. Harari believes evidence of this lies in the fact that there is no archaeological evidence of trade at Neanderthal sites. Each Neanderthal group would have manufactured its own tools from local materials. In contrast, 30,000-year-old Homo sapiens sites in inland Europe have been found to contain shells from the Mediterranean. It is highly likely that these shells reached the continental interior through long-distance trade between different tribes of Homo sapiens [1].

Symbolic culture: Non-material culture, including concepts, beliefs, institutions and language, normally considered to be constructed uniquely by Homo sapiens.

Harari notes that trade cannot exist without trust, and if Sapiens traded shells, they would also have been able to trade information, creating a much wider knowledge network than previous human species. The ability to transmit knowledge and new behaviours without the need for genetic adaptation would have triggered a revolution of progress. This is not to say that the laws of nature no longer applied to Sapiens. According to Harari, "Our physical, emotional and cognitive abilities are still shaped by our DNA. Our societies are built from the same building blocks as Neanderthal or chimpanzee societies, and the more we examine these building blocks, the less difference we find between us and other apes" [1].

Another type of cooperation that influenced language was the habit of rearing offspring collectively. Because ape mothers bring up their young alone, this may explain why apes did not increase their repertoire of sounds to the same extent as humans did, nor did they develop the art of dialogue. According to Judith Burkart, Professor of Anthropology at the University of Zurich, raising young collectively requires a lot of communication. The offspring have to make sure that they get enough food and attention, so they constantly make noises to interact with their caregivers. Those who are better at maintaining contact increase their chances of survival. This, in turn, is likely to lead to the formation of new neural networks that support communicative behaviour [2].

Professor Burkart has been studying the pro-social behaviour of marmosets, the tiny New World monkeys native to South America. Unlike the great apes, where mothers raise their young alone, marmoset mothers and fathers take turns looking after their offspring. Such a form of collective rearing is not found in other non-human primates; it is only otherwise practised by humans. Burkart claims that because of their pro-social behaviour, marmosets communicate extensively and as a result use very complex language, considering their small brains. They use combinations of sounds and are masters at sharing information. They also start calling back and forth as soon as they are separated from each other, and in a way that resembles human dialogue. This remarkable parallel has made marmosets an important object of study in anthropology, and their pro-social behaviour is the central argument for explaining why humans and apes have evolved differently in terms of their cognitive abilities and their language [2] (see also the "social complexity hypothesis" in Sect. 3.3).

Burkart also points out that if a mother takes care of her infant alone and has to wait each time until it is independent, she can only have a few offspring in her lifetime. This slows down the rate of reproduction and in extreme cases can lead to the extinction of a species. Humans solved this problem by having the entire family (including siblings, aunts and grandmothers) take care of the young, and it is this pro-social behaviour that enabled humans to develop and pass on ever more complex language [2].

Thus, adaptation to savannah life necessitated cooperation and communication not only in procuring food but also in raising offspring, which in turn influenced the development of human language. Up to this point, we have explained language largely in terms of an adaptive feature, driven by natural selection, in response to the physical and social environment of human beings. However, since Darwin's time, it has been argued that the human language capacity has reached a level of complexity that far exceeds what could be achieved by natural selection alone [3]. The difficulties posed by the relatively sudden emergence of language are often referred to as "Darwin's problem". Noam Chomsky famously avoided the issue of selection by attributing the language capacity to a random genetic mutation, but we can also turn to a very different explanation of how language developed.

5.2   The Cultural Evolution of Language

In recent years, researchers have begun to appeal to the cultural evolution of language to explain the origins of structure in language, providing an alternative to nativist and adaptationist accounts of language evolution. Language, like other cultural systems, spreads across the community via social interaction (in a process known as **cultural diffusion**) and is passed on from generation to generation via children's language learning (in a process known as **cultural transmission**). Over time, this produces small changes in language as certain structures are abandoned in favour of others, leading to the **cultural evolution** of language. According to British linguist Kenny Smith, "As long as language has been culturally transmitted, cultural processes and cultural evolution will have been at work" [4]. It follows that any capacity to use and acquire language cannot be considered in isolation from the dynamics of cultural transmission and cultural evolution. Smith toys with the idea that cultural evolution could potentially offer a uniform mechanism that explains "both the genesis of language (a qualitative shift from a non-linguistic system to a linguistic system) and language change (subsequent quantitative shift)". He nevertheless notes, "Even if we subscribe to the strongest cultural account, we must still explain the emergence of the capacity for a culturally transmitted system of communication" [4].

Cultural diffusion: Process by which culture, including language, spreads gradually across a community through social interaction.

Cultural transmission: Process by which culture, including language, is passed on from one generation to the next through imitation and learning.

Cultural evolution: Language change through social interaction and cultural transmission rather than genetic inheritance.

5.3   An interplay of Culture and Biology

Terrence Deacon, Professor of Anthropology and Neuroscience at the University of California, Berkeley, approaches the problem by suggesting there is a complex interaction between biological and cultural evolution, leading in the long term to what is known as **gene-culture co-evolution**. This is in line with "**niche construction**" theories in biology (e.g. Odling-Smee et al. 2003 [5]) that suggest that the changes organisms make to their environment subsequently cause them to adapt to the new environment. Deacon explains that the emergence of symbolic language created a totally new kind of niche for humans which, in turn, demanded a new set of adaptive features. He notes that just as beaver dam-building has produced an aquatic niche to which beavers' bodies and behaviours have adjusted in the course of evolution, so our cognitive capacities have evolved to our self-constructed **symbolic niche**. According to Deacon, the need to link an ever-increasing number of concepts to an ever-increasing number of symbols would have favoured features such as the ability to learn by mimicking, complex memory systems and superior vocal control [6].

Gene-culture co-evolution: Theory that proposes culture plays an active role in the evolution of genes.

Niche construction: Process in biology whereby an organism alters its local environment, thus interfering with natural selection and changing gene frequencies in populations.

Symbolic niche: Self-constructed environment consisting of symbolic behaviour to which humans adapt.

5.4 Language Self-evolving

Assuming that a more complex mode of communication was critical for integration into human social groups and that it contributed to successful reproduction, selection would have favoured all traits that enabled this enhanced form of communication. However, we must not forget that language evolution includes an additional twist. As language spreads across a community and is transmitted from one generation to the next, it gets reshaped by its users and learners. In the words of Deacon, "Language itself exhibits an evolutionary dynamic that proceeds irrespective of human biological evolution" [3]. Attempts have been made to demonstrate these processes using computer simulations that model the evolution of language, e.g. de Boer's simulation of the emergence of vowel systems [7]. Using a population of "artificial agents" that can perceive and produce vowels in a human-like way, de Boer demonstrated that the vowel systems that emerge from simulations bear a close resemblance to human vowel systems (which are optimal for communication). He concluded that this optimisation is the result of **self-organisation** and that innate biological predispositions are probably not the best explanation for the universal tendencies of human vowel systems.

Self-organisation: Idea that the structure and patterns in language are not genetically pre-programmed but arise from multiple pressures acting on language as it emerges and changes in socially interacting populations.

Some even see language change as a Darwinian process. The very interesting **"cultural selection theory"** models cultural change on theories of biological evolution. Like Darwin's natural selection theory, cultural selection theory has three phases: variation, reproduction and selection; variation leads to a new form; reproduction accounts for the spread, and selection controls the spread [8].

Cultural selection theory: Theory that models cultural change on Darwinian natural selection.

According to science philosopher Gillian Crozier [9], the cultural selection theory is an extension of **memetics**, which proposes that "memes" (coined in Richard Dawkins's 1976 book *The Selfish Gene*), much like genes in biology, are autonomous units of information hosted in the minds of individuals and passed through generations of culture. Linguistic memes, named "linguemes" by Croft [10], are "a population of very diverse kinds of units: phonological features, phonemes, syllables, rules of phoneme sequencing, lexicons, rules of syntax, semantic categories, systems of world categorisation, constructions mapping combinations of words and complex meanings, prosodic structures, social conventions involving gestures and gaze to coordinate linguistic interactions, etc." [11]. Some critics have argued that memes do not exist independently but depend on a physical embodiment in some medium; most, however, would agree that the theory is based on analogy and simply an approach to cultural information transfer.

Memetics: Theory of information transfer proposed by Richard Dawkins, according to which memes, much like genes in biology, are units of information passed through generations of culture.

According to cognitive and computational scientists Oudeyer and Kaplan [11], in the last fifty years or so, language has been largely regarded as a fixed and idealised system, accounted for by universal principles of language structure, where individual variation is ignored or left unsolved. If, on the other hand, language is viewed as a system of replicators, with variation as a central concept, this changes the emphasis:

language is no longer a static or pre-designed system, but rather a dynamic and adaptable one that evolves over time [11]. Cognitive scientists Morten Christiansen and Nick Chater [12] have suggested that language evolves and adapts to the cognitive constraints and **learning biases** of its users and learners, i.e. linguistic structure is shaped by how language is processed and learnt. This means that for language to survive, it must firstly be easy to articulate and comprehend. Secondly, since language has to be learnt, the more learnable its structures are, the more effective its reproduction in each generation will be.

Learning biases: Innate preferences and tendencies that individuals have when acquiring a language. For example, language learners are sensitive to regularities in language, which helps them identify linguistic patterns and structures.

Another way in which language is considered to be shaped is by the communicative needs of its users. According to evolutionary linguist Dr. Christine Cuskley [13], successful language must firstly be useful: it must express the many meanings its users need to communicate. Secondly, language needs to have expressive power: it must be able to integrate new forms into existing structures. I would add here a third condition: language must be efficient in its task of communicating messages, for example, by reducing redundancy while maintaining precision.

The notion that language self-evolves was visibly witnessed when groups of deaf children who had been denied access to sign language were brought together in a school in Nicaragua after a change of regime in the 1980s. The children were not taught sign language, but one evolved spontaneously (**Nicaraguan Sign Language**), much as a **pidgin** language develops among people who do not have a language in common. After ten years or so, the next generation of pupils took on the new sign language but made their own grammatical innovations (as happens with a **creole** language), which the previous generation did not adopt. For example, the sentence

The woman	tapped	the man
S	V	O

was progressively signed in a more complex way. The first generation signed it as follows:

woman	tap	man	get tapped
N	V	N	V
(sign for woman)	(sign for tap made into the air)	(sign for man)	(sign for tap made on the shoulder)

The second generation used the location of the sign to distinguish the subject and object:

woman	tap	man	get tapped
S	V	O	V
(sign for woman made to the left)	(sign for tap made into the air)	(sign for man made to the right)	(sign for tap made on the shoulder)

The third generation brought in a further innovation. Instead of producing the sign for "woman" to the left and the sign for "man" to the right, both signs were produced centrally followed by pointing to the left or right. This phenomenon demonstrates that language has a natural drive to complexity, but that only younger learners have the ability to innovate and create new structures. The Nicaraguan experience also shows that language emerges from social interaction and the need to connect with one another.

> **Nicaraguan Sign Language (NSL)**: Sign language that developed spontaneously among deaf children in a school in Nicaragua in the 1980s.

> **Pidgin**: Simplified means of communication that develops among people who do not have a language in common.

> **Creole**: Pidgin that has become a group's first language and therefore undergoes dramatic expansion.

British linguist Simon Kirby [14] suggests that the structure of language arises from the interactions of three complex systems: biological evolution, individual learning biases and cultural transmission. He explains that as individuals, we acquire language using learning mechanisms that are part of our biological endowment. This "learning machinery" is the mechanism by which language is transmitted culturally through a population of individuals over time. Ultimately, this process of cultural transmission results in a set of language structures that bring fitness advantages to individuals using those structures. This, in turn, leads to the biological evolution of language learners, closing the loop of interactions (see Fig. 5.1).

This model rejects neither the theory of cultural evolution of language nor the theory of a biological foundation for language. Instead, it stresses co-dependence: the cultural medium of language simultaneously depends on and shapes the biological capacity to learn language from others.

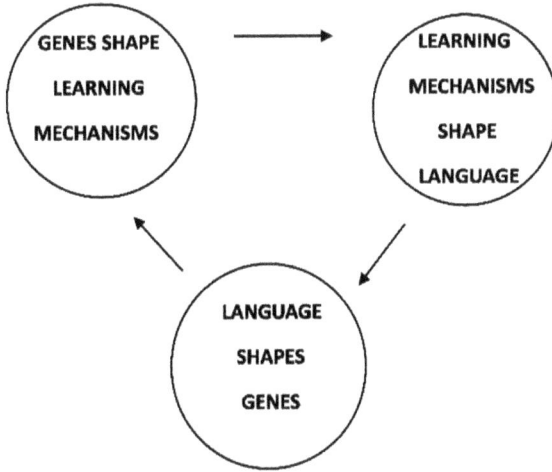

Fig. 5.1 Kirby's loop of three adaptive systems involved in the emergence of language (adapted from Kirby et al. 2007: Fig. 1) [14]

I will finish the chapter with a further example of an **agent-based model**, this time one that simulates the emergence of linguistic structure through "chains of learning". The influential "iterated learning model" [15] starts off with agents learning a subset of a randomly produced artificial vocabulary. The resulting output of this first generation of agents becomes the new input for the next generation. Since agents receive only a subset of the input, they apply any emerging patterns to all other items and, as a result, patterns accumulate in the vocabulary over the successive generations of learners. This line of research emphasises the significance of cultural transmission in language evolution. While each individual agent has only a small bias towards structure, this bias is greatly amplified by transmission, producing a highly structured language. These approaches highlight the key role that cultural processes can play in creating linguistic structure [14].

Agent-based models: Computational models that simulate the actions and interactions of autonomous agents (e.g. language users) in order to understand the behaviour of a system (e.g. language).

5.5   Chapter Summary

In this chapter, we saw how our complex language is the result of a variety of social factors, including the need for cooperation and social cohesion, a prolonged childhood and a complex social organisation. In addition, we saw that not only humans have adapted to communicate more effectively but language itself has evolved to become more usable and learnable through the processes of cultural transmission and cultural evolution. Finally, we saw that the structure of language can be seen as arising from the interactions of three complex systems: biological evolution, individual learning and cultural transmission.

In the next chapter, we will take a closer look at the processes involved in historical language change and see how a cultural evolution process might fit into the larger picture of language evolution.

References

1. Harari, Y. N. (2014). *Sapiens: A brief history of humankind*. Harvill Secker.
2. Gull, T. (2021). Mensch und Schimpans. *UZH Magazin, 2*(21), 32–37.
3. Deacon, T. (2010). A role for relaxed selection in the evolution of the language capacity. *PNAS, 107*(2), 9000–9006. https://doi.org/10.1073/pnas.0914624107
4. Smith, K. (2006). Cultural evolution of language. In E.-CÂÂ. K. Brown (Ed.), *Encyclopedia of Language & Linguistics* (2nd ed., pp. 315–322)
5. Odling-Smee, J., Laland, K. N., & Feldman, M. W. (2003). Niche construction: The neglected process in evolution. *Monographs in Population Biology, 37*. https://doi.org/10.1515/978140 0847266
6. Deacon, T. (1997). *The symbolic species: The co-evolution of language and the brain*. Norton.
7. de Boer, B. (2000). Self organization in vowel systems. *Journal of Phonetics, 28*(4), 441–465. https://doi.org/10.1006/jpho.2000.0125
8. Wikipedia. (2023). *Cultural selection theory*. https://en.wikipedia.org/wiki/Cultural_select ion_theory
9. Crozier, G. K. D. (2008). Reconsidering cultural selection theory. *The British Journal for the Philosophy of Science, 59*(3), 455–479. https://doi.org/10.1093/bjps/axn018
10. Croft, W. (2000). *Explaining language change*. Longman.
11. Oudeyer, P., & Kaplan, F. (2007). Language evolution as a Darwinian process: Computational studies. *Cognitive Processing, 8*, 21–35. https://doi.org/10.1007/s10339-006-0158-3
12. Christiansen, M., & Chater, N. (2008). Language as shaped by the brain. *Behavioral and Brain Sciences, 31*, 489–558. https://doi.org/10.1017/S0140525X08004998
13. Cuskley, C. (2020). *Language evolution: A brief overview*. PsyArXiv. https://doi.org/10.31234/ osf.io/3y98j
14. Kirby, S., Dowman, M., & Griffiths, T. L. (2007). Innateness and culture in the evolution of language. *PNAS, 102*(12), 5241–5245. https://doi.org/10.1073/pnas.0608222104
15. Kirby, S., & Hurford, J. R. (2002). The emergence of linguistic structure: An overview of the iterated learning model. In A. Cangelosi & D. Parisi (Eds.), *Simulating the evolution of language* (pp. 121–147). Springer.

Chapter 6
The Role of Historical Language Change in the Evolution of Language

One of the aims of evolutionary linguistics is to discover how historical language change (language change over historical time) fits into the larger scheme of language evolution. At one extreme, authors such as Berwick and Chomsky [1] assume language change merely constitutes surface modifications to the biological **faculty of language (FL)**, while authors at the other extreme, such as Heine and Kuteva [2], believe language change is a crucial mechanism in the emergence of human language. Before we can resolve this issue, we need to consider what exactly language change is and how is it caused.

> **Faculty of language (FL)**: Uniquely human cognitive system proposed by Noam Chomsky that supports the acquisition and use of language.

For Spanish linguist José-Luis Mendívil-Giró at the University of Zaragoza, such conflicting views of language change presuppose a very different notion of what is changing when language changes. He notes, "According to the biological model, language evolution is a matter of the biological evolution of organisms … that happens on a geological timescale (in the order of hundreds of thousands of years)" [3]. An underlying assumption is that natural evolution first creates variation by genetic mutations which are inherited by future generations. It should be mentioned here that patterns of inheritance are not only affected by natural selection. Changes in a population's gene pool can also be the result of random genetic drift (when certain individuals leave more offspring purely by chance) or gene flow (caused by individuals migrating from one population to another). Having said that, the biological model of language evolution rests heavily on Darwin's adaptive natural selection, where the differential survival and reproduction rates of individuals with favourable traits lead to the gradual accumulation of beneficial characteristics within populations over time. Some biological accounts of language evolution focus on the

J. Dornbierer-Stuart, *The Origins of Language*,
https://doi.org/10.1007/978-3-031-54938-0_6

anatomical and physiological adaptations for language, while others focus on the cognitive.

According to the "cultural" model, historical language change is the key mechanism in the evolution of language. It argues that the faculty of language does not exist as such—it is not coded in the genes. Rather, grammatical structure arises through **language use**. Historical linguists Bernd Heine and Tania Kuteva [2] postulate that grammar is created by the process of **grammaticalisation**, where, during the course of time, lexical items (such as nouns and verbs) become "grammaticalised" and develop into grammatical items (such as auxiliaries, case markers, inflections and prepositions). For example, it is well known that the phrase "let's", now functioning as an auxiliary to introduce a suggestion, evolved from the lexical cluster "let us", meaning "allow us" [4]. According to Heine and Kuteva, all that is needed for grammaticalisation to take place is "a linguistic system that (a) is used regularly and frequently within a community of speakers and (b) is passed from one group of speakers to others" [2].

Language use: Use of language in concrete situations for communication. Accounts for considerable variation of language over time.

Grammaticalisation: Process of language change whereby lexical items (nouns and verbs) develop over time into grammatical items such as auxiliaries, case markers, inflections and prepositions. Believed by some to be a key mechanism in the origin of grammar.

So how do historical linguists obtain their evidence for such claims? We will now turn to the methods that can be used to measure change in language, using both the diachronic and synchronic approach.

6.1   Measuring Language Change

The modern scientific study of language change over time (diachronic or **historical linguistics**) grew out of the earlier disciplines of philology (the deciphering and study of ancient written texts and documents) and etymology (the study of the source and development of present-day words). By the nineteenth century, it had become popular to compare different languages and hypothesise about their common ancestry (**comparative historical linguistics**). The comparative method was initially applied to Indo-European languages but later also to languages outside of the Indo-European family. The first step involved finding cognates (systematic sound and meaning corre-spondences) between attested languages. For example, within the Polynesian group, we find the following sound correspondences in the numerical system (Table 6.1).

By plotting similarities between existing languages, it was possible to hypothesise parent languages from which all the languages had descended (e.g. Proto-Polynesian) and, through further **linguistic reconstruction**, even hypothetical ancestors of these parent languages (e.g. Proto-Austronesian). However, any reconstructed system will be far from perfect as many words that appear related may have initially been imported from other languages instead of developing directly from the parent language.

> **Historical linguistics**: Study of language change over time and the development of theories to explain the underlying processes.

> **Comparative historical linguistics**: Comparative study of languages for evidence of ancestral proto-languages.

> **Linguistic reconstruction**: Method used to reconstruct earlier forms of a language by using evidence from correspondences in later languages.

Up until the mid-twentieth century it was deemed impossible to observe change in progress, since it was assumed that change could only be registered once it had occurred. Then, around fifty years ago, a breakthrough was achieved with the recognition that all the synchronic variations in a language (e.g. different pronunciations) are an important source of language change and an indication that change is happening. We will explore the concept of **linguistic variation** later in the chapter, but first, we will use evidence from a thousand years of English preserved in writing to determine the types of linguistic changes that take place.

Table 6.1 Cognates among languages of the Polynesian family

	One	Two	Three	Four	Five
Tongan	taha	ua	tolu	fā	nima
Samoan	tasi	lua	tolu	fā	lima
Māori	tahi	rua	toru	ɸā	rima
Hawaiian	kahi	lua	kolu	hā	lima

> **Linguistic variation**: Different ways of saying the same thing. Variation occurs in all levels of grammar and can be attributed to physiological, psychological and sociological factors.

6.2   Types of Change

Old English texts from the tenth century such as the epic poem *Beowulf* reveal just how much English has changed, to the extent that Old English is now virtually unintelligible to the modern speaker of English. On closer examination, we can see there have been changes in all parts of grammar, including phonology, morphology, syntax, the lexicon and semantics, which we will now explore.

6.2.1   Phonological Change

Phonemes can change over time. For example, the k in *elk*, now pronounced as a voiceless velar stop /k/ in Modern English, used to be pronounced as a voiceless velar fricative /x/ (as in Scottish *loch*) in Old English. Many sound changes are conditioned by the phonetic environment and occur uniformly across the lexicon. For example, other words where the final /x/ sound in Old English evolved into the /k/ sound in Modern English are *milk* and *book*. Sometimes, sound changes take hold in a single word and then similar looking words follow suit through a process of analogy. This gradual diffusion of a particular sound change through the lexicon is known as **lexical diffusion**. In addition to sounds changing, phonemes can also be lost or gained. For example, the gh in *night, light, right* and *bright* was once pronounced /x/ but became silent in Late Middle English, whilst the phoneme /ʒ/ was added to the sounds of Early Modern English with the borrowing of French words such as *rouge* and *beige* [5].

> **Lexical diffusion**: Gradual spread of a particular sound change to all similar-looking words in the vocabulary of a language.

Between 1400 and 1600, there was a major change in the pronunciation of English vowels, known as the **Great Vowel Shift**. The seven long vowels of Middle English /ɑː ɛː eː iː ɔː oː uː/ underwent an increase in tongue height, with the result that *goose*, once pronounced /goːs/, became /guːs/, and *geese*, once pronounced /geːs/, became /giːs/. Today, this shift has left traces in certain morphemes that have two different pronunciations, e.g. the /iː/ in *please* /pliːz/ reverts back to /e/ in *pleasant* /ˈplezənt/

Most languages also adopt words from other languages, known as **lexical borrowing**. It is well known that after the Norman Conquest, many words infiltrated the English language from Old French (e.g. *baron, peasant, government, priest*), and later from Middle and Modern French (e.g. *chef, hotel, etiquette, avant-garde*). Between 1500 and 1700, English imported many learned words from Greek and Latin. In addition, English has words from Norse (e.g. *sky, skin*), Celtic (e.g. *clan, bin*), Dutch (e.g. *buoy, leak*), German (e.g. *quartz, beer*), Italian (e.g. *opera, piano*), Arabic (e.g. *algebra, zero*) and Spanish (e.g. *guitar, ranch*) [5]. In addition to words being gained in a language, many words fall into disuse. You probably haven't heard the word *snollygoster* for a long while.

Lexical borrowing: Importing words into a language from other languages.

6.2.4 Semantic Change

It is also common for lexical items to shift in meaning. A word may become broader in meaning (e.g. *holiday* changed its meaning from "a day off work for worship" to "a day off work"), or else narrower (e.g. *deer*, from German *Tier*, once meant "beast" and now refers to a particular kind of animal). A word may also change its meaning more radically (e.g. *silly* used to mean "happy" in Old English, "naïve" in Middle English, only to become "foolish" in Modern English) [5].

6.3 Mechanisms of Change

Having seen what sort of changes occur in language, we will now look at what causes language to change. As mentioned previously, there are various forces acting on language as it is transmitted "horizontally" across a community (through social interaction) and "vertically" from one generation to the next (through children's language learning). In addition, language change can occur due to movements of population and internal changes in population size. We will start with the effects of language transmission.

6.3.1 Horizontal Transmission

Horizontal transmission (otherwise known as cultural diffusion) results in "intra-generational" linguistic changes that occur as a consequence of "using" language.

There are various mechanisms at work, influenced by both internal and external factors. The first set of changes are "internally" motivated—they are related to phys-iological factors (such as ease of articulation) and psychological factors (such as the tendency to simplify structure). Starting with the physiological, some sounds and combinations are easier to pronounce than others. The **"theory of least effort"** suggests sound changes are primarily due to linguistic laziness. For example, few people today pronounce the d in *handbag* or *handkerchief*, and no one pronounces the t in words such as *whistle, thistle* and *castle* [7].

Horizontal transmission: Transmission of language across a community through social interaction.

Theory of least effort: Theory that suggests the sounds in words change to facilitate articulation.

Tok Pisin (formally a pidgin spoken throughout Papua New Guinea that now has native speakers and so has gained the status of a creole) has undergone extremely rapid sound change. The increased fluency brought about by native speakers has caused assimilation and elision of speech sounds, leading to changes such as:

man bilong mi (my husband) → *mamblomi*

American linguist Joan Hooper (now Bybee) found that phonological reduction affects high-frequency words earlier than low-frequency words, as seen with deletion of schwa /ə/ in *memory* (/'meməri/ → /'memri/) but not in *mammary* (= /'maməri/) [8]. Moving on to psychological factors, it is also human nature to look for inherent patterns in language and tidy up irregularities. For example, the Middle English word for "pea" was *pease,* but it was gradually assumed that *pease* was plural, and a new singular *pea* came into being [7]. At a similar time, the past of *catch* became *caught,* following the analogy of *teach-taught,* just as today the past of *sneak* is becoming *snuck,* in accordance with *stick-stuck.*

The second set of changes are "externally" (socially) motivated. American linguist William Croft proposes two primary mechanisms here. First, language users intro-duce innovations into the linguistic system, creating variants (e.g. *singin'* instead of *singing*). Second, due to **social prestige**, certain variants are chosen in preference to alternative forms and spread through the community. According to Croft, this model, involving innovation and socially motivated propagation, potentially provides a mechanism for both the emergence and subsequent evolution of a linguistic system [9].

Social prestige: In sociolinguistics, the high regard gained by using a specific language form or variety within a speech community. Prestige can be accorded to both standard forms of a language (overt prestige) and non-standard forms (covert prestige).

Pioneering sociolinguist William Labov is famous for his fieldwork in socially motivated language change. He observed that lower status groups tend to imitate higher status groups. He also observed that women are more susceptible to social pressure from "above" (i.e. women are more likely to adopt prestige forms) and men are more susceptible to social pressure from "below" (i.e. men are more likely to adopt stigmatised forms) but that, on the whole, women are more likely to lead language change by using more innovative forms than men.

Since Labov's work, other models of socially motivated language change have been developed. In addition to social prestige, there are other motivations for **intraspeaker variation** (individuals' style-shifting): one has to do with "accommodation" and another with identity building. In Allan Bell's "Audience Design" model [10], speakers accommodate their language to their addressee, whereby there are two processes: **convergence**, in which speakers adapt their language towards that of the interlocuter to achieve social acceptance, and **divergence**, in which speakers distance themselves from the interlocuter by steering their language away from that of the interlocuter. In Penelope Eckert's "Speaker Design" model [11], speakers use variables to associate and dissociate themselves with others, thus constructing a social identity. With all of these models (involving prestige, accommodation and identity building), speakers are using language variation to create social meaning in their daily interactions.

Intraspeaker variation: Individuals' style-shifting, causing language variation and change over time.

Convergence: Process in sociolinguistics whereby speakers adapt their language towards that of the interlocuter to achieve social acceptance.

Divergence: Process in sociolinguistics whereby speakers distance themselves from the interlocuter by steering their speech away from that of the interlocuter.

6.3.2   Vertical Transmission

Vertical transmission (otherwise known as cultural transmission) leads to "intergenerational" linguistic changes that occur due to children's language learning. In the process of language acquisition, each child constructs their own unique grammar, generalising rules from the linguistic input they receive. The child's grammar is never exactly like that of the adult generation. On the one hand, "external" factors are to blame; children receive variable input from many different styles used in different social settings from which to decipher the norms. On the other hand, "internal" factors, such as limited memory and attention (see Chapter 7), mean that children tend towards simpler structures. As a result, linguistic structures that are highly regular will be more likely to survive cultural transmission intact than structures that are irregular. Then again, frequently expressed concepts will be more easily memorised and under less pressure to become regular. This is why the ten most common verbs in English (*be, have, do, say, make, go, take, come, see, get*) have all kept their irregular past tense forms. Children may initially regularise certain common irregulars (e.g. *goed, taked, sheeps, mouses*), although errors of this type normally get ironed out at a young age.

Vertical transmission: Transmission of language from parents to offspring through imitation and learning.

But children do not only "simplify" structure; they may also "create" structure that is absent in adult language. This was seen nowhere more clearly than in the spontaneously developing Nicaraguan Sign Language (introduced in Chapter 5). Each successive generation of learners brought innovations to the language which the older generation speakers did not adopt. This experience revealed the vital role children play in language change, both in the innovation of structure and in its adoption. Moreover, by observing the linguistic processes involved in NSL's development, researchers were able to gain insights into the fundamental mechanisms underlying the emergence and development of a new language, with obvious implications for the origin of human language.

"Real-time" studies of spoken language change over several generations are for practical reasons difficult and rare. An effective alternative involves the "apparent time" construct, where a linguistic variable (a linguistic feature with more than one realisation, such as the ending "-ing" in *singing*) is studied in speakers of different ages at a single point in time. In this way, synchronic data is used to simulate change over time. Empirical research comparing the language of different age groups within a single-speaking community indicates that adolescents play a prominent role in introducing and adopting new forms, no doubt reflecting their desire to differentiate themselves from parents (divergence) and forge commonality among peers (convergence). Adult speakers, too, although more conservative in their language, contribute to linguistic change through their interactions, occupations and exposure to new ideas and language varieties. Additionally, other social groups, such as ethnic minority communities, professionals, academics or individuals with specific interests, may also drive linguistic change in particular domains or subcultures. Thus, linguistic change is a far-reaching and complex phenomenon that is determined by both internal (physiological and psychological) and external (social and cultural) factors.

6.3.3 Demographic Processes

In addition to the internal and external forces acting on language, as it is transmitted horizontally across a community and vertically from one generation to the next, there are demographic processes driving language change. These include (i) patterns of migration and (ii) internal population changes. As for the former, immigration can cause one language or language variety to merge with another, and emigration can cause languages to split and develop in new directions. As for the latter, increases in population size can cause new group oppositions with corresponding linguistic innovations. Ultimately, all these processes need to be incorporated into a theory of language evolution.

We started off the chapter by distinguishing between a biological model of language evolution that attempts to specify a "language capacity" common to all humans, and a cultural model that tries to account for language change and the great diversity of languages in the world. It is often assumed there is a clear-cut distinction between the emergence of language and subsequent language change, but according to Dediu et al. [12], it might be more profitable to assume a continuous development throughout the whole history of language, with differences in degree but not in quality. Both biology and culture are sources of variation, and change is ongoing in both. Biological change might proceed more slowly than cultural evolution, but there is interaction between the two.

In their recent survey of new directions in language evolution research, Nölle et al. [13] note that we already have a variety of methods from comparative biology, palaeontology, neuroscience and genetics to study the biological components of language. Likewise, there are multiple experiments from the lab to address the cultural evolution of language. With the aid of computer modelling, more recent studies are now able to cross disciplinary boundaries to integrate the biological and cultural aspects of language, with the result that language is coming to be viewed more holistically. According to the researchers, what we need is a theory where all interacting factors are causally linked, especially with regards to how language is shaped by language-internal and external environmental factors. Nevertheless, they conclude that the focus of language evolution research today has already made the shift towards "uncovering the co-evolutionary dynamics of language, cognition and culture" [13].

6.4   Chapter Summary

We started by looking at exactly what changes when language changes. We saw changes in all parts of grammar, including phonology, morphology, syntax, the lexicon and semantics. Then we moved on to what causes language to change. Firstly, we distinguished between horizontal transmission, the cause of intra-generational change, and vertical transmission, the cause of inter-generational change. Within horizontal transmission, we saw various mechanisms of change relating to either internal (processing) factors (e.g. ease of articulation) or external (social) factors (e.g. social prestige). Likewise, change due to vertical transmission is influenced by both internal factors (e.g. the limited memory of infants) and external factors (e.g. the variable input from adults). Lastly, we looked at demographic processes causing language change, including patterns of migration and internal population changes. By the end of the chapter, it was clear that language change arises due to an interplay of biological, social, demographic and cultural processes, all of which cannot be ignored in our account of language evolution.

In the next chapter, we will turn to the subjects of psycholinguistics and neurolinguistics to illuminate the cognitive and neural mechanisms behind being able to produce and understand language.

References

1. Berwick, R.C., & Chomsky, N. (2016). *Why only us: Language and evolution.* MIT Press.
2. Heine, B., & Kuteva, T. (2007). *The genesis of grammar.* Oxford University Press.
3. Mendívil-Giró, J.-L. (2019). Did language evolve through language change? On language change, language evolution and grammaticalization theory. *Glossa: A Journal of General Linguistics, 4*(1), 124. 1–30. https://doi.org/105334/gigl.895
4. Wikipedia. (2023). *Grammaticalisation.* https://en.wikipedia.org/wiki/Grammaticalization
5. Fromkin, V., & Rodman, R. (1983). *An introduction to language* (3rd ed.). CBS College Publishing.
6. Meyerhoff, M. (2018). *Introducing sociolinguistics* (3rd ed.). Routledge.
7. Aitchison, J. (1991). *Language change: Progress or decay?* (2nd ed.). Cambridge University Press.
8. Hooper, J. B. (1976). Word frequency in lexical diffusion and the source of morphophonological change. In W. Christie (Ed.), *Current progress in historical linguistics* (pp. 96–105). North Holland.
9. Croft, W. (2000). *Explaining language change: An evolutionary approach.* Longman.
10. Bell, A. (1984). Language style as audience design. *Language in Society, 13*, 145–204. https://doi.org/10.1017/S004740450001037X
11. Eckert, P. (2003). *The meaning of style.* Stanford University.
12. Dediu, D. et al. (2013). Cultural evolution of language. In P. J. Richerson & M. H. Christiansen (Eds.), *Cultural evolution: Society, technology, language and religion* (Vol. 12, Ch. 16, pp. 303–332).
13. Nölle, J., Hartmann, S., & Tinits, P. (2020). Language evolution research in the year 2020. *Language Dynamics and Change, 10*(2020), 3–26. https://doi.org/10.1163/22105832-bja10005

Chapter 7
How do we Produce and Understand Speech?

A comprehensive model of language evolution should not only describe the structure of the communication system that has evolved but also define the mechanisms underlying its use. In this chapter, we shall therefore enter the realm of **psycholinguistics** and **speech processing** and look at the mechanisms involved in generating speech (speech production) and understanding the speech of someone else (speech perception). Although these two processes are often treated separately, they are, as we shall see, inextricably linked. The chapter will focus on models of speech processing, before turning to some evidence from neuroscience to add to our understanding of how language is processed in the brain.

Psycholinguistics: Discipline that studies and tests theories of how language is processed and represented in the mind. It uses a number of non-invasive techniques to discover how we produce, comprehend and acquire language.

Speech processing: Sensory, motor and cognitive processes involved in understanding and producing spoken language.

In the past, psycholinguists gained insights into speech processes using two methods. The first method was to study the speech of aphasics (people who have a linguistic deficit, or **aphasia**, following damage to a part of the brain involved in language). As we saw in Chapter 4, aphasic individuals often exhibit selective impairments in one of the broad categories of speech production (Broca's aphasia) or language comprehension (Wernicke's aphasia), and this helps to identify the brain regions critical for these functions. More specific impairments include disruptions in

J. Dornbierer-Stuart, *The Origins of Language*,
https://doi.org/10.1007/978-3-031-54938-0_7

discriminating or producing sounds, word retrieval, syntactic and semantic integration, understanding figurative language and auditory–motor integration. Analysing specific types of errors can offer valuable clues about the underlying cognitive processes and the neural mechanisms involved in different aspects of language.

The second method used by psycholinguists was to observe slips of the tongue, or **speech errors**, in healthy individuals. This type of evidence in particular suggests that our brains process language in a modular and organised way. For instance, slips of the tongue that involve substitutions (e.g. *foon speeding* instead of *spoon feeding*) supply evidence that we recognise and deal with discrete linguistic units when forming words; the substitution of "f" for "sp" in *foon speeding* demonstrates that speakers have distinct phonemes (or combinations of phonemes) stored in their mental inventory, and the substitution of "speeding" for "feeding" indicates that speakers recognise the morphemic structure of the words. Studying these errors can help us understand the mechanisms behind speech in the human brain, which could provide clues about how these mechanisms evolved.

Aphasia: Inability to produce fluent speech (expressive or Broca's aphasia) or meaningful speech (receptive or Wernicke's aphasia) because of damage to specific brain regions.

Speech error: Unconscious deviation from the intended form of an utterance, otherwise known as a slip of the tongue.

More recently, these insights have been complemented by the findings of **neurolinguistics**, which uses neuroimaging to illuminate the neural underpinnings of language. Both approaches (psycholinguistic and neurolinguistic) support the view that language consists of interacting modules, which will influence our view of how language evolved.

Neurolinguistics: Discipline that deals with the relationship between language and the structure and functioning of the brain; historically rooted in the study of linguistic deficits (aphasias) occurring as the result of brain damage.

7.1 Speech Production

Insights from speech error research were central to the first models of **speech production**, which attempt to represent the process by which thoughts are transformed into coherent speech. Most models incorporate a number of operations, including conceptualisation, linguistic encoding and articulation. Conceptualisation involves abstractly identifying the proposition to be expressed; linguistic encoding involves transforming the proposition into a linguistic form, and articulation involves physically producing the speech signal. Additionally, models often incorporate forward-planning mechanisms, a buffer (working memory storage) and a monitoring mechanism [1].

Speech production: Process whereby thoughts are transformed into coherent speech. Involves conceptualisation, linguistic encoding and articulation.

According to Professor Hans-Jörg Schmid at LMU Munich, central controversies in speech production revolve around the notion of **modularity**. The first issue concerns whether the different stages in speech production are handled by different, highly specialised modules in the brain, and, if so, whether these modules are strictly separated or closely interlinked. He notes, "The fact that there are some stroke patients who have fluent grammar but difficulties finding the right words and some who know all the words they need but cannot assemble them in grammatical sentences is evidence for some sort of neurological specialisation in the brain, since these specific deficits can be correlated with the regions in the brain that are damaged" [2].

Modularity: Idea that a system is composed of independent modules that interact and function as a whole. In linguistics, the term is used in two ways, either to refer to the faculty of language as a distinct module of the mind, or else to refer to the fact that the human language capacity is composed of various modules that handle different aspects of language.

A second issue concerns whether these modules work in parallel or sequentially. For example, are syntactic frames and words selected simultaneously, or does the output of one module serve as input for the next one? [2] In the highly influential model of speech production proposed by Dutch psycholinguist Willem Levelt [3], the "conceptualiser" provides input for the "formulator", which provides a framework for the "articulator". Figure 7.1 shows a simplified version of the model.

In the first stage of the model (**conceptualisation**), the concept that the speaker wishes to convey is selected from an array of semantically related concepts. This generates a pre-verbal message which serves as input for the next component, the formulator.

Fig. 7.1 Processing
components of speech
production (adapted from
Levelt 1989) [3]

```
┌─────────────────────────────────────────┐
│                                         │
│            CONCEPTUALISER               │
│                                         │
│                                         │
│      Generation of pre-verbal message   │
│                                         │
└─────────────────────────────────────────┘
                    ↓

┌─────────────────────────────────────────┐
│                                         │
│              FORMULATOR                 │
│                                         │
│                                         │
│           Linguistic encoding           │
│                                         │
└─────────────────────────────────────────┘
                    ↓

┌─────────────────────────────────────────┐
│                                         │
│              ARTICULATOR                │
│                                         │
│                                         │
└─────────────────────────────────────────┘
                    ↓
                 Speech
```

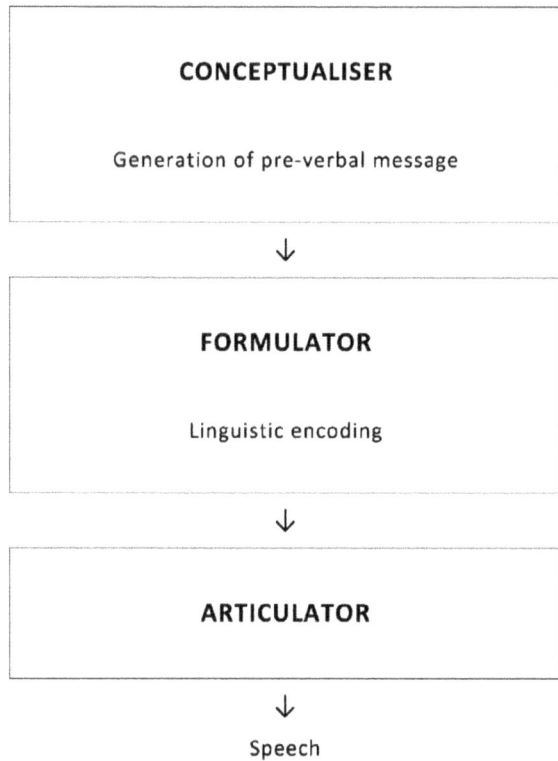

> **Conceptualisation**: In speech processing, the initial stage in speech production involves the selection of the concepts to be expressed and identification of the relationships between them.

The formulation stage (**linguistic encoding**) involves transforming the message into a linguistic form at various levels: grammatically, morphologically, phonologically and phonetically. Grammatical encoding is the process of selecting appropriate lemmas (root words), e.g. "he", "have", "two", "house", from our mental lexicon (see Sect. 2.5) and ordering them according to the syntactic rules of the language (he have two house). Each word retrieved from the mental lexicon has conditions regarding the syntactic environment it requires. These environments will be different for verbs, nouns, adjectives and prepositions. Morpho-phonological encoding is the process of determining the correct inflectional forms, turning lemmas into lexemes, e.g. have → has, house → houses, and breaking these down into syllables: he-has-two-hou-ses. The final part of the formulation stage is phonetic encoding, which involves computing the phonological shape (sound) of the syllables and creating an articulatory score ready for spell-out. The precise sounds within a syllable will depend on the sounds of the preceding and subsequent syllables.

> **Linguistic encoding**: Process in speech production whereby thoughts are transformed into a linguistic form with phonological shape, ready for articulation.

The third component of speech production is **articulation**, which involves executing the articulatory score using the lungs, vocal cords, tongue, lips, jaw and other parts of the vocal apparatus involved in speech. A whole network of neural substrates in the cerebral cortex, basal ganglia and cerebellum are involved in this, forming our most complex motor behaviour [4].

> **Articulation**: Final stage in speech production involving the physical production of an acoustic speech signal.

Following this scheme, speaking is based on a sequence of processing components that are relatively autonomous in their functioning, but where each module sets up the necessary conditions for the next module to function. While Levelt's model provides valuable insights into the cognitive processes underlying speech production, recent advances in neuroscience have suggested a more complex and distributed organisation of language in the brain.

However it happens, it truly is an amazing feat for anyone to be able to coordinate, virtually instantaneously, all the components necessary to produce even a single word, and it is even more astounding that a baby manages to articulate its first recognisable word at around the age of twelve months. To produce the typical first word *mama* requires tremendous coordination since it involves all of the following: rapidly drawing in an appropriate breath of air, slowing down exhalation, controlling the vocal cords to produce a buzzing at an appropriate pitch, joining the lips in the correct way while keeping the soft palate open to produce the nasal "m" sound, keeping the centre of the tongue relatively low in the mouth to produce the "a" sound, doing all of that twice over and knowing that the noise you have made refers exclusively to that very special person in your life! However, being able to generate speech is only half the story. Let us turn now to speech perception, the process by which the sounds of language are heard, interpreted and understood.

7.2 Speech Perception

Speech, the end product of speech production, is the starting point for **speech perception**. Speech emanating from the speaker's articulatory apparatus is received by the listener's auditory apparatus. Reception of the sound signal is accompanied by **acoustic processing**, which involves breaking down a continuous stream of speech into meaningful elements (words and phrases) as it arrives in real time ("bottom up" processing). Then, using higher-level language processes connected with morphology, syntax and semantics, these elements are fit into a precon-ceived grammatical structure that has been learned and stored in the memory ("top down" processing). The phonetic strings are assigned linguistic structure (**linguistic decoding**) via the "parsing component" and eventually transformed into messages with meaning via the "conceptualiser".

Speech perception: Process whereby the sounds of language are heard, interpreted and understood. Involves acoustic processing, linguistic decoding and semantic interpretation.

Acoustic processing: Initial stage in speech perception whereby acoustic features such as pitch, intensity and duration are detected and used to break down the speech signal into meaningful elements.

Linguistic decoding: Process in speech perception whereby grammatical information is extracted from a stream of speech in order to arrive at the intended message conveyed by the speaker.

Figure 7.2 is a simplified version of Levelt's 1993 model of spoken language use [5], including speech production and speech perception. It highlights the bi-directionality of the two processes. Although not exact reversals, the processes are closely intertwined: both processes rely on the same linguistic knowledge (phonology, morphology, syntax) and work together to facilitate effective commu-nication.

Speech perception is a task that humans carry out effortlessly, yet it involves accessing and coordinating many different types of information—not only segmental and suprasegmental phonological information but also syntactic and semantic infor-mation—all within milliseconds. Acoustic processing presents the first hurdle for several reasons. Firstly, the input sounds are produced very rapidly. In normal speech, we are dealing with between ten and thirty distinct speech sounds per second. Secondly, the boundaries between words are very indistinct. In a speech spectro-gram (a photographic representation of the sound waves in speech), there are no blank spaces between the words of continuous speech. Try saying *This is an apple* with slight pauses between the words and you will see how unnatural/robotic it sounds. A third complication is that speech sounds are modified by the other sounds around them. Sounds at the ends and beginnings of words may be swallowed up, making word boundaries even more indistinct. For example, "Have you told him?" in natural speech often becomes "V'you told'im?" Even within words, sounds can

```
┌─────────────────────────────────────────────────────────┐
│                    CONCEPTUALISER                         │
│   Message generation          Message interpretation      │
└─────────────────────────────────────────────────────────┘
         ↓                              ↑
┌──────────────────────┐      ┌──────────────────────┐
│     FORMULATOR        │      │       PARSER          │
│  Linguistic encoding  │      │  Linguistic decoding  │
└──────────────────────┘      └──────────────────────┘
         ↓                              ↑
┌──────────────────────┐      ┌──────────────────────┐
│     ARTICULATOR       │      │  ACOUSTIC PROCESSOR   │
└──────────────────────┘      └──────────────────────┘
         ↓                              ↑
┌─────────────────────────────────────────────────────────┐
│                        SPEECH                             │
└─────────────────────────────────────────────────────────┘
```

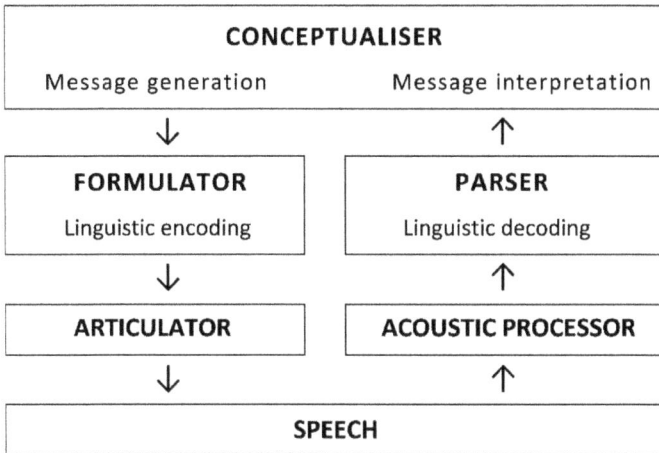

Fig. 7.2 Processing components in spoken language use (adapted from Levelt 1993) [5]

change into different sounds. If you say "Edinburgh" quickly, you will probably pronounce it as "Edimburgh", and "inquiry" as "ingquiry".

Yet, to distinguish the meaning of words, it is essential to recognise the sounds in words. If this process is disturbed, the result is dyslexia. Problems of sound recognition can lead to difficulties later on when it comes to learning to read and write because when we start reading, our brain translates the letters into sounds. If we are already unable to recognise and categorise sounds correctly, this makes reading (and life) difficult for us.

7.3 The Role of Attention and Working Memory

The first neural models of speech perception, (e.g. Wernicke, 1874 [6]; Lichtheim, 1885 [7]) assumed "passive" processing, whereby auditory inputs project directly to perceptual interpretations without involving higher cognitive processes or feedback loops. Later models incorporated "active" processes, acknowledging the recruitment of cognitive resources like attention and working memory. **Attention** is an important mechanism that enables us to highlight relevant features of input data while excluding irrelevant features. Since the amount of data the brain can process at once is limited, the attention mechanism allows information to be prioritised for processing. **Working memory** is a cognitive system that holds information temporarily to help the brain make sense of it. It puts the information into some kind of order. Once the information has been processed, it exits the brain's short-term repository and passes into the long-term memory for future retrieval.

> **Attention**: Cognitive mechanism that enables us to highlight relevant features of input data. It allows information to be prioritised for speech processing.

> **Working memory**: Dynamic cognitive system that holds information temporarily through the persistent firing of neurons. Important for reasoning and decision-making as well as speech processing.

Experiments have shown that when we listen to language, we use context (involving working memory) to predict upcoming items. For example, consider how the context provided at the start of this sentence speeds up the processing of the final part:

They wanted to make the hotel look more like a tropical resort.

So, along the driveway, they planted rows of ... [8]

Probably you came up with *palms* rather than *pines*. Working memory not only plays a role in predicting upcoming items during speech perception but also facilitates the forward planning of responses during speech production. The processes involved in turn-taking are an integral part of human language and require cognitive skills which would have developed through specialisation.

7.4   The Motor Theory of Speech Perception

Historically, research in speech processing has been split into the two distinct fields of speech perception and speech production, not least because methodologies are necessarily quite different. Research is now focusing on the inherent links between these two processes. Since the signals generated in speech production are the same as those received in speech perception, the two processes must at some point be dealing in the same linguistic currency.

It was the "motor theory of speech perception" proposed by the American behavioural psychologist Alvin Liberman in 1967 that first suggested a "meshing" of the perception and production of speech in the brain. The theory argues that the speech motor system is not only involved in producing speech articulations but also in detecting them [9]. More precisely, people perceive spoken words by identifying the vocal gestures with which they are articulated rather than the sound patterns they generate. With the paradigm shift in experimental psychology from behaviourism to cognitive psychology, the hypothesis gradually lost credibility but was revived in the early 1990s by the discovery of **mirror neurons**. A team of neuroscientists led by Giacomo Rizzolatti found that there are specific neurons in the motor cortex of macaque monkeys that are activated when both observing an action and performing an action [10]. There is also fairly strong evidence from neuroscience (e.g. Rizzolatti and Arbib, 1998 [11]) that speech perception activates the same areas of the brain involved in the articulation of speech. This coupling between the perceptual system and the motor system forms the basis of **vocal imitation**, which seems to be a crucial mechanism behind babies' first words. According to the theory, as babies watch their parents articulate sounds, motor stimulation of the articulators aids recognition of the sounds. Without this coupling system, there would be no language learning, and no language.

Mirror neurons: A type of neuron that is thought to cause a perceived action to be simulated in the perceiver's brain. Considered to form the basis for imitation and social learning.

Vocal imitation: Mechanism whereby an individual observes and replicates another's vocal behaviour. Sensory experience is internalised and used to shape vocal outputs.

It has also been proposed that the motor system is involved in the process of **symbol grounding** (linking linguistic symbols to meaning) by linking mental concepts with sensorimotor experience. According to this view, when we observe a horse galloping, and at the same time we hear the word "horse", an association is forged between the visual experience of the horse and the auditory experience of the word "horse". Hearing the word "horse" again in another situation reactivates the visual experience, and this reactivation of the visual experience enables us to understand the word, since it is "grounded" in sensorimotor experience [12]. However, most words are actually learnt in a linguistic context without any direct contact with their referents (e.g. *uranium, freedom*), i.e. they are learnt in a symbolic fashion, via their relations to other symbols. Nevertheless, it seems likely that involvement of the motor system, a concept known as **embodiment**, is a necessary first step in linking symbols to meaning, whether in children's language learning or humans' first words.

Symbol grounding: Cognitive process by which abstract concepts and symbols are connected to real-world sensory experience through the engagement of the motor system. Crucial for understanding and using language effectively.

Embodiment: Use of the motor system to link mental concepts with sensory experience.

7.5   Some Neurolinguistic Research

I will conclude the chapter with a couple of examples of how research in neuroscience has contributed to psycholinguistic models of speech processing. The first example involves determining the precise timing of phonological, syntactic and semantic processes during speech perception. We have already seen that data from patients with brain lesions can indicate what kind of brain structure is necessary to support what part of language (e.g. syntax or semantics). This builds a picture of the neuroanatomy of speech (see speech areas of the brain in Sect. 4.4). In addition to this, we can use electroencephalography **(EEG)** to capture the electrical activity of the brain, millisecond by millisecond, along with event-related brain potentials **(ERPs)**, to determine the order and time course of different processes (e.g. syntactic versus semantic) during speech processing.

Electroencephalography (EEG): Method used for recording the electrical activity of the brain. Electrodes placed on the scalp are linked to an electroencephalograph, which produces visible brainwave patterns.

Event-related potentials (ERPs): Changes in the brain's electrical activity in response to specific stimuli. In the context of language processing, the so-called N100 (peaking at 100 milliseconds) has been linked to initial acoustic processing, the ELAN (peaking at 200 milliseconds) to syntactic structure building and the N400 (peaking at 400 milliseconds) to semantic processing and thematic relations assignment.

In 2002, neuropsychologist Angela Friederici at the Max Planck Institute for Human Cognitive and Brain Sciences in Leipzig proposed a neurocognitive model of speech comprehension based on these two types of data. As a starting point, Friederici specified the functional neuroanatomy of speech comprehension as follows (see Table 7.1).

Using EEG data (collected from healthy individuals while they listened to carefully crafted sentences with semantic or grammatical errors) combined with ERPs, Friederici suggested the following distinct phases in speech comprehension (see Table 7.2).

Table 7.1 The functional neuroanatomy of speech comprehension

	Temporal regions	Frontal regions
Left hemisphere	Identification of phonetic, lexical and structural elements	Sequencing and formation of structural, semantic and thematic relations
Right hemisphere	Identification of prosodic parameters	Processing of sentence melody

Based on Friederici (2002) [13]

Table 7.2 Neurocognitive model of speech comprehension in the left hemisphere

Phase 0 (0–100 ms)	Primary acoustic analysis and identification of phonemes and word forms
Phase 1 (100–300 ms)	Formation of initial syntactic structure, based on information about word category
Phase 2 (300–500 ms)	Lexical-semantic processing, including thematic role assignment (agent, patient, etc.)
Phase 3 (500–1000 ms)	Integration of the different types of information

Adapted from Friederici (2002: Fig. 1) [13]

In short, Friederici argues that the building of syntactic structure precedes semantic processes, although both processes interact during a later phase. She notes that prosodic processes (occurring predominantly in the right hemisphere) interact with syntactic processes, but that the exact timing of this is a subject for future research [13]. Friederici's model still holds today, but she believes that we now need to focus attention on speech processing at the molecular level. She claims that understanding how individual neurons and synapses function will not only help us understand how language works, but also how it evolved in human beings. In her 2017 book *Language in our Brain,* foreworded by Noam Chomsky, she shows how different language-related brain regions are connected to make language possible, and she compares the findings in humans with other primate species. She claims that one particular fibre bundle of the dorsal tract, associated with the processing of syntactically complex sequences, is only weakly present in non-human primates and could therefore be at the root of the human capacity for language.

The second example of how neuroscience can contribute to speech processing models involves comparing the brain connections in speakers of two very different languages during speech production. Researchers Wei et al. (2023) [14] at the Max Planck Institute in Leipzig have recently demonstrated that native speakers of German (a morpho-syntactically complex Indo-European language) exhibit stronger structural connectivity within the syntax network of the brain, while speakers of Arabic (a root-based Semitic language) exhibit stronger connectivity within the lexical-semantic network. These differences are not genetically determined, but instead reflect the brain's remarkable capacity to accommodate different linguistic environments effectively. This adaptive capacity, known as **neuroplasticity**, is a core feature of the brain which, over evolutionary time, has of course been favoured by natural selection as an advantageous trait.

Neuroplasticity: Ability of neural networks in the brain to respond and adjust to environmental influences, which has implications for learning new skills. The developing brain of a child exhibits a higher degree of plasticity than the adult brain.

Finally, it should be noted that communication does not merely consist of a simple linguistic encoding/decoding algorithm shared by speaker and hearer, but also encompasses a huge amount of inference from the non-linguistic environment: "It's hot in here!" may actually mean "Can someone open the window?" The speaker's utterance can rarely be taken word for word but must be interpreted against the broader communicative context in order to arrive at the speaker's actual intention. In the last decade, brain imaging techniques have also been increasingly combined with EEG data to investigate pragmatic processing, but the process of meaning construction is notoriously hard to specify since it is based on a highly developed set of cognitive abilities for social cognition. These abilities nevertheless constitute a vital component of human language and thus of any theory of language origins.

7.6   Chapter Summary

In this chapter, we looked into the various components involved in speech perception (including acoustic processing, linguistic decoding and comprehension) and speech production (including conceptualisation, linguistic encoding and articulation). In addition, we saw how the two processes are inextricably linked, especially when it comes to the mechanisms of imitation and symbol grounding, which are crucial in children's language learning and no doubt also in humans' first words. We also saw how the findings of neuroscience can better inform psychological models of speech processing. By combining insights from both disciplines, researchers can develop more comprehensive and accurate frameworks that account for both the cognitive processes and their underlying neural mechanisms. This interdisciplinary approach facilitates a deeper understanding of how the brain supports speech and language functions. Since the 1990s, brain imaging techniques have been increasingly used to investigate phonological, grammatical and semantic processes and, more recently, prosodic and pragmatic processes, as well as their interaction, in the brain. We still struggle to understand the overall brain architecture supporting pragmatics, since it involves pathways between brain areas implicated in linguistic processing and those implicated in non-linguistic functions (such as theory of mind). This all needs to be accounted for in a theory of language origins.

In the next chapter, we will see what clues ontogeny (the development of language in children) can provide for phylogeny (the emergence of language in the human species). We will explore the early stages of children's language acquisition, starting with mastering the sound system, through to producing first words and simple sentences, and see how these stages might relate to a theory of language origins.

References

1. Field, J. (2004). *Psycholinguistics*. Routledge.
2. Schmid, H.-J. (2012). *Linguistic theories, approaches and methods*. In M. Middeke, T. Müller, C. Wald, & H. Zapf (Eds.), *English and American studies*. J. B. Metzler.
3. Levelt, W. (1989). *Speaking: From intention to articulation*. The MIT Press.
4. Levelt, W. (1995). The ability to speak: From intentions to spoken words. *European Review, 3*(1), 13–23. https://doi.org/10.1017/S1062798700001290
5. Levelt, W. (1993). The architecture of normal spoken language use. In G. Blanken, J. Dittman, H. Grimm, J. C. Marshall, & C.-W. Wallesch (Eds.), *Linguistic disorders and pathologies: An international handbook* (pp. 1–15). Walter de Gruyter.
6. Wernicke, C. (1874). *Der aphasische Symptomencomplex: eine psychologische Studie auf anatomischer Basis*. Cohn & Weigert.
7. Lichtheim, L. (1885). Über Aphasie. *Deutsches Archiv für klinische Medizin, 36*, 204–268.
8. Federmeier, K. D., & Kutas, M. (1999). A rose by any other name: Long-term memory structure and sentence processing. *Journal of Memory and Language, 41*, 469–495. https://doi.org/10.1006/JMLA.1999.2660
9. Liberman, A. M., Cooper, F. S., Shankweiler, D. P., & Studdert-Kennedy, M. (1967). Perception of the speech code. *Psychological Review, 74*(6), 431–461. https://doi.org/10.1037/h0020279
10. di Pellegrino, G., Fadiga, L., Fogassi, L., Gallese, V., & Rizzolatti, G. (1992). Understanding motor events: A neurophysiological study. *Experimental Brain Research, 91*, 176–180. https://doi.org/10.1007/BF00230027
11. Rizzolatti, G., & Arbib, M. A. (1998). Language within our grasp. *Trends in Neurosciences, 21*(5), 188–194. https://doi.org/10.1016/s0166-2236(98)01260-0
12. Günther, F., Dudschig, C., & Kaup, B. (2017). Symbol grounding without direct experience. *Cognitive Science, 42*(2), 336–374. https://doi.org/10.1111/cogs.12549
13. Friederici, A. D. (2002). Towards a neural basis of auditory sentence processing. *Trends in Cognitive Sciences, 6*(2). https://doi.org/10.1016/S1364-6613(00)01839-8
14. Wei, X., Adamson, H., Schwendemann, M., Goucha, T., Friederici, A. D., & Anwander, A. (2023). *Native language differences in the structural connectome of the human brain*. Max Planck Institute for Human Cognitive and Brain Science, Department of Neuropsychology, Leipzig, Germany.

Chapter 8
How do we Learn to Speak?

In addition to speech processing, a second concern of psycholinguistics is how language develops in childhood. In this chapter, we will examine the cognitive capacities and conditions required for children's language acquisition and see if these can be applied to a theory of how language emerged in the human species. It is an enticing notion that children's language development may in some way mirror the early development of human language. Just as children progress from babbling and single words to complex sentences, so too could protolanguage have made developmental steps that spread from generation to generation. The two processes obviously differ in that children from the modern era are exposed to a fully developed language system from the outset. Nevertheless, studying the order in which mental capacities develop in infants can tell us about the cognitive structures and mechanisms on which language builds, and thus contribute to the language evolution discussion. The chapter begins with a description of the developmental stages of language learning in children and uses these as a basis for determining how much of language can be attributed to an innate capacity, and how much depends on the social environment.

All humans, unless socially isolated as children or suffering from certain physical or mental impairments, become accomplished speakers of at least one language within a few years of childhood and with little discernible effort. Although there is considerable variation in the timing and duration of the different phases of language development in individual children, the sequence of development is fairly uniform across all languages. Table 8.1 shows the main stages, with approximate timing.

The consistency across languages would seem to suggest that language development is very much dependent on "internal" (biological) constraints. On the one hand, it is clear that the younger child has cognitive limitations such as limited memory and attention, which will affect when what can be learnt. On the other hand, we know from studies in developmental psychology that babies do not enter the world as "blank slates" but have a predisposition to find patterns in the environment and learn certain information early on in life. This predisposition makes children active, self-motivated learners practically from birth.

© The Author(s), under exclusive license to Springer Nature Switzerland AG 2024   97
J. Dornbierer-Stuart, *The Origins of Language*,
https://doi.org/10.1007/978-3-031-54938-0_8

Table 8.1 The main stages of language development in children

	Linguistic Capacity	Focus
1 month	Able to distinguish different sounds	Sound System
3 months	Production of first speech sounds	
6 months	Babbling begins	
9 months	Intonation patterns emerge	
12 months	First words	Words and Meanings
20 months	Two-word utterances (without inflections)	Syntax and Communicative Development
24 months	Inflections; three-word utterances	
30 months	Questions; negatives; four-word utterances	
4 years	Correct but limited syntax	
5 years	Complex syntax	

For example, at a very early stage of development, human babies respond to the melody and rhythm of speech. It is this inborn musical ability that initially enables infants to make sense of the cacophony of speech. A little later, babies are able to distinguish between different speech sounds, determine which sounds may be combined with which in their particular language and assign meaning to the sound combinations. In order to produce speech sounds and first words, the ability to imitate is crucial. As we saw in Chapter 7, the simultaneous involvement of auditory and visual perception and motor action helps in linking words to meaning and memorising the connection. However, all of these innate language learning mechanisms cannot come into play without social interaction and exposure to the language being learnt and are therefore dependent as much on the "external" environment as on "internal" constraints. We will now take a closer look at each stage of children's language development.

8.1 0–12 Months: Learning the Sound System

The first stage in children's language acquisition centres on developing awareness of the language's sound system. Studies have shown that almost from birth babies pay attention to the source of sounds and within a few days can discriminate the human voice from other sounds [1]. At one month, babies can distinguish between consonant pairs such as [b] and [p]. This can be demonstrated, for example, by repeatedly playing the sound [ba] while infants are sucking a dummy. When the sound changes to [pa], there is an abrupt increase in the rate of sucking. This inborn ability to distinguish between meaningful elements (known as **categorical perception**) is also found in other species. For example, songbirds have been shown to categorically perceive variations in the pitch or timing of song elements. This suggests that the ability to create categories in language depends on perceptual mechanisms that existed prior to the emergence of humans.

Categorical perception: Innate ability to perceive categories along a continuum (e.g. colours along the colour spectrum, or phonemes along the sound spectrum). In human language, boundaries between categories are determined largely by convention.

Production of speech sounds follows a little later. In the first few weeks, babies' vocal sounds are simply responses to physical and emotional states such as pain, hunger or surprise, equivalent to the symptomatic calls of other species. At around three months, babies have more control over their speech organs and start to produce their first speech sounds. We have already seen (in Sect. 7.4) that imitation, involving cross-modal coordination, plays a crucial role at this point. In addition to imitation, there is also a certain amount of experimentation, which is considered by many to be an essential precursor to vocal learning. During **babbling**, it is not only the sounds that are practised but also intonation patterns. Some have suggested that the different intonation contours corresponding to different emotions (such as pleasure and distress) are among the first linguistic contrasts that children perceive.

Babbling: The stage in children's language acquisition during which an infant appears to be experimenting with producing sounds. Also occurs in some animals and songbirds.

Infants' early sensitivity to intonation is well documented. Mehler et al. [2] discovered that at just four days old, infants can distinguish the intonation patterns of their native language from those of another language, and Mampe et al. [3] found that the cries of four-day-old infants already correspond to the intonation contours of their mother tongue. Friederici et al. [4] found that infants of two months can distinguish between long and short syllables. In addition, Friedrich et al. [5] found that there was a correlation between two-month-old infants with a family risk of a speech disorder and deviant brain responses to long and short syllables. In other words, a speech disorder can potentially be recognised in infants as young as two months.

Studies of slightly older infants have revealed sensitivity to both the accenting in words (important for recognising words) and stress patterns in whole sentences (which help in identifying syntactic phrasing). For example, Friederici et al. [6] compared the processing of two-syllable words in four-month-old German and French infants. The researchers created a pseudoword *baba*, stressed either on the first syllable (*bába*), as is typical in German, or on the second syllable (*babá*), as is typical in French. In a first round, they presented the infants with *bába* as the standard and *babá* as a deviant, and in a second round, they switched this around. Using event-related potentials (ERPs), they found a different brain activity pattern in the two groups. The data for German infants demonstrated a clear positive mismatch response only when the deviant stimulus had French stress, but not when the deviant stimulus had German stress, while the data for French infants revealed exactly the opposite. The researchers concluded that each language group displayed a processing advantage for the rhythmic structure typical of their native language, indicating "language-specific neural representations of word forms in the infant brain as early as four months of age" [6]. Pannekamp et al. [7] found children of eight months recognised intonational phrase boundaries, revealing prosodic processing at the sentence level.

Evidence of children mastering whole sentence patterns before they can articulate words can be clearly observed in the babbling dialogue of twins. It is often reported that twins appear to be having "grown-up" conversations, with questions, answers, facial expressions and gestures, before they can produce any proper words. In rare cases, twins develop their own "secret" language that only they can understand. They actually make up words that they both use.

This rare glimpse of child language development could be seen to represent the "cusp of language" stage in the origin of language. As we saw in Chapter 3, the analytic model of language evolution suggests that humans, like human infants, may have first produced holistic musical cadences which were later segmented into smaller meaningful parts (words). However, the fact that speech melody is dealt with primarily by the right hemisphere of the human brain, while mainstream language systems (phonology, syntax and semantics) are processed in the left hemisphere implies that the latter probably did not develop out of the former but that both developed simultaneously. According to British linguist Maggie Tallerman, the two systems are rather distinct. She compares human holistic vocalisations (like laughing and sobbing) to the more primitive calls of non-human primates, which are largely involuntary and genetically encoded, i.e. they are not learnt. Speech, on the other hand, is under voluntary control and culturally transmitted [8]. Nevertheless, it cannot be denied that musicality is a biological sub-component of speech, without which human language may never have taken off.

8.2   12–20 Months: Proto-Words and First Words

During a transitional period between babbling and the first word, children produce **proto-words**, i.e. invented words that are used consistently to express certain meanings, e.g. *Eh!* used as a request. The next milestone in children's language acquisition is at around 12 to 14 months, when the first recognisable words are produced. First words tend to be short and are often based on the CV (consonant + vowel) pattern, for example *boo* for "boot" or *dada* for "daddy". Most first words refer to objects, but some refer to actions (*look*), some are modifiers (*up*) and some are terms of social interaction (hello). As with proto-words, these first words are often used to express a whole message. For instance, *Down!* may be a request to put something down, or a statement that the child has sat down. These meaningful words are known as "**holophrases**". It may well be that the first words of humans, being limited in number, were used to signal a whole situation, before grammar came onto the scene.

> **Proto-word**: A very early word-like utterance produced by an infant (e.g. *Eh!* used as a request) before true first words appear. Can also refer to a word-like utterance produced by early humans before they had the capacity for full language.

> **Holophrase**: Single word expressing the whole idea in a phrase or sentence, e.g. *Down!* to mean "I want to get down".

This seems to be the case with evolving pidgin and creole languages (see Chapter 5), which tend to have small vocabularies, but each word has a broader range of meaning. In Tok Pisin, a creole language spoken throughout Papua New Guinea, *pisin* (from English *pigeon*) means "bird in general", and *gras* (from English *grass*) can signify "grass", but also "hair", "fur" or "feathers". Pidgin languages are illuminating as they show how humans naturally devise a simple language and thus give clues as to how human language might have evolved.

At around 18 months, there is a rapid increase in vocabulary production. Some have suggested that the first true words coincide with the **naming insight**, the moment when children realise things in the world can be labelled. Some believe that the naming insight was the defining moment that distinguished human language from other animal communication systems, when humans acquired "some specific knowledge about the symbolic properties of language" [9]. Although monkeys can use different alarm calls to signal different types of danger, these calls are linked to here-and-now situations involving survival and are largely genetically predetermined. Monkeys cannot take the signal for a threatening snake and put it in another sentence such as "there are no threatening snakes around today". Only humans can "displace" signals to refer to something that is not physically present. Subsequent to children's first words, vocabulary growth is rapid and can be put down to the "snowball effect" or cumulative learning.

> **Naming insight**: In language acquisition, the moment when a child realises that things have labels, whether the things are present or not. Some believe the naming insight was a crucial moment in the evolution of human language.

According to Friederici [10], lexical-semantic processing can be neurally demonstrated in infants of 14 months by using the picture-word interference task. This involves infants being shown a series of pictures of different objects. For each object they see, they simultaneously hear a word for that object. If there is a mismatch between the object they see and the word they hear, brain activation patterns will change. More recent research, e.g. Bergelson and Aslin (2017) [11], is showing that already at six months, well before first words appear, babies are constructing cross-word relations. For example, when faced with two images of common nouns, babies look significantly more at the named target image when the competitor image is semantically unrelated (e.g. milk and foot) than when it is related (e.g. milk and juice). This suggests that semantic relatedness between visual objects influences initial word comprehension in babies.

As for the beginnings of human language, this suggests that semantic structure is a fundamental component of language. According to Jackendoff [12], it is the first "generative" (creative) component of language: the conceptual system provides the underlying concepts and abstract relationships that are then mapped onto the specific words and grammatical structures of a particular language. This is in direct contrast to Chomsky's theory of generative grammar, which assumes syntax is the only generative component of language onto which sound and meaning are subsequently mapped. These opposing views of language will be revisited within the evolutionary context in Chapter 10.

A final observation during this stage, at 15–20 months, is that children transitioning from the one-word stage to the two-word stage use pointing or iconic gesturing (see Sect. 3.3) to indicate subject-predicate constructions. For example, *Daddy* (uttered while pointing to a plant) could signify "Daddy watered the plant", and *hair* (uttered while gesturing hair washing) might signify "I washed my hair". Since this type of deliberate gesturing can be observed in apes, it seems possible that intentional vocalisations might have developed from gestures [13]. Some have suggested that gestural control "collatoralised" speech control, since gestures and speech are processed in neighbouring regions of the motor cortex. According to Arbib et al. [14], "The fact that aphasia of signed, and not just spoken, languages may result from lesions based on Broca's area supports the view that one should associate Broca's area with multimodal language production rather than with speech alone" [14]. However, gestural hypotheses of language origins cannot account for the well-developed phonology found in speech. It seems that gesturing, like musicality, is another type of scaffolding that language is built upon.

8.3   20–30 Months: Simple Grammar

The next recognisable stage in children's language development starts at around twenty months, when two words are combined. This is a very significant step as it represents the beginning of syntax, or **proto-syntax**. There is much more going on than simply saying one word after another: stress and intonation patterns change according to the syntactic function of the phrase or clause. For example, one common early syntactic structure is the "agent-action" combination, where a child may say something like *Mummy eat* to express the idea that Mummy is eating. In this case, the intonation pattern would involve ascending pitch on the first word (↗*Mummy*) and descending pitch on the second word (↘*eat*). *Mummy coat* would have a different stress depending on whether it meant "Mummy is putting on her coat" ("coat" is stressed) or "That's Mummy's coat" ("Mummy" is stressed). Generally, at this stage, only "content" words are used—mainly nouns and verbs—while the "function" words—mainly articles, prepositions and auxiliary verbs—are left out, and there are no inflections (endings) indicating number, person or tense, resulting in what is known as **telegraphic speech**. Again, musicality is key in communicating meaning.

Proto-syntax: The earliest form of syntax produced by infants, featuring two-word utterances. Can also refer to the earliest form of syntax produced by humans, whereby two words were joined sequentially to signal more than one idea.

Telegraphic speech: Two-word utterances representing simple grammar, e.g. *Get down* to mean "I want to get down".

When children start to produce utterances of more than two words, at around 24 months, content words still predominate but the utterances already exhibit the word order and intonation patterns of adult sentences. For example:

Adult phrase:	*John's*	*got*	*two*	*cups*	*of*	*juice*
Infant phrase:	*John*	*got*	*two*			*juice*

For correct intonation, there must be an "understanding" of how words are grouped into phrases and phrases into sentences. **Phrase structure grammar** is a formal method employed by linguists to represent this knowledge. It uses an upside-down tree-like structure to break down sentences into their constituent parts. The grammar works "top down": it starts with the sentence, which is broken down into the subject (a noun phrase) and predicate (a verb phrase), which in turn are broken down into smaller phrases and eventually words. The sentence *John's got two cups of juice* is analysed as in Fig. 8.1.

Phrase structure grammar: Formal method employed by linguists to break down sentences hierarchically into phrases and words using an upside-down tree-like structure.

Phrase structure grammar uses the same subject-predicate division of Latin and Greek grammar, which can be traced back to Aristotle's ontological categories and parts of speech. In fact, Aristotle's linguistic principles can be compared to those of modern **cognitive linguistics**, which assumes that linguistic structures closely reflect the way we perceive and experience the world. Aristotle divided the clause into two

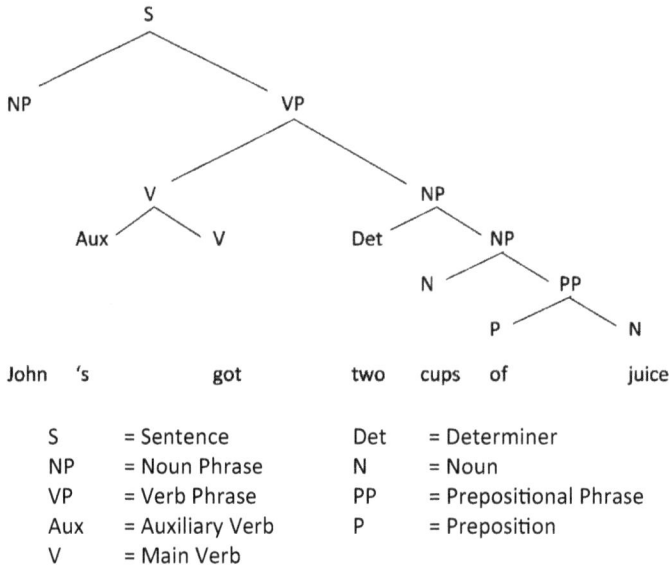

```
                              S
                    ┌─────────┴─────────┐
                   NP                    VP
                                ┌────────┴────────┐
                                V                  NP
                            ┌───┴───┐        ┌─────┴─────┐
                          Aux       V       Det          NP
                                                      ┌───┴───┐
                                                      N       PP
                                                          ┌───┴───┐
                                                          P       N

        John    's          got         two      cups   of        juice
```

S	= Sentence	Det	= Determiner
NP	= Noun Phrase	N	= Noun
VP	= Verb Phrase	PP	= Prepositional Phrase
Aux	= Auxiliary Verb	P	= Preposition
V	= Main Verb		

Fig. 8.1 Sentence analysis using phrase structure grammar

parts: the subject (roughly speaking, "who" or "what" is doing an action) and the predicate (roughly speaking, what action the subject is doing—the verb—and to whom—the object). While the precise order of subject, verb and object varies from language to language, the relative position of these constituents is an important way of conveying causal relationships in a sentence.

Cognitive linguistics: Movement in linguistics that grew as a reaction to generative grammar. A fundamental tenet of cognitive linguistics is that language is not a separate module of the mind but is deeply intertwined with general cognitive capacities such as perception, attention, memory and conceptualisation.

Researchers Hirsh-Pasek and Golinkoff [15] used elicitation techniques to demonstrate that even one-word speakers aged 16–19 months recognise the significance of word order in the sentences that they hear. For example, in the "preferential looking technique", children were presented with two videos, side by side, one where Cookie Monster is tickling Big Bird and the other where Big Bird is tickling Cookie Monster. The children were then asked to find Big Bird tickling Cookie Monster, i.e. they had to look at the video where Big Bird is doing the tickling rather than Cookie Monster. Yuan and Fisher [16] demonstrated that by 27 months, toddlers can learn novel verbs for actions that are not currently observable simply by hearing them in conversations and using word order and other syntactic cues to derive their meaning.

How is all this relevant to a theory of language origins? Putting strings of words together in a set order would not only have aided language learning in early humans but would have also made communication more effective. In a social context, having a consistent word order helps convey intentions and meanings more clearly, which would have been crucial for cooperation and coordination among early human groups. As humans evolved and culture developed, the need to express more complex ideas, relationships and distinctions would have led to an expansion of grammar. This is exactly what happens in the next stage of children's language acquisition.

8.4 Beyond 30 Months: Complex Grammar

A common view is that grammar is constructed incrementally by using general learning strategies, based on a mix of problem-solving and a process of trial and error. Just as the meanings of words may be overgeneralised during early language learning (e.g. *Mummy* may refer to all females), so too can the rules of grammar. For example, many English children produce the form *hisself* rather than *himself* (following the pattern *myself*, *yourself*) [1]. The tendency to regularise is also evident in plural formations like *mouses/mices* and *foots/feets*, and in verb forms like *comed* and *goed* [1]. Over time, children's grammar will eventually align to adult patterns. On the other hand, various studies (commencing with Brown, 1973 [17]) have uncovered a striking consistency in the route by which grammatical structures are mastered. For example, negative constructions in English are initially formed by tacking the negative to the front of the utterance (*No sun shining*). Later, *no* is inserted into the clause (*Sun no shining*). Finally, auxiliaries are mastered (*The sun isn't shining*) [1]. This seems to suggest that language learning is innately guided and that language structure is, so to speak, hardwired into the brain. Let us look at each of these contrasting perspectives in turn, starting with the "constructionist" view of language learning.

According to the pioneering developmental psychologist Michael Tomasello in his influential book *Constructing a Language: A Usage-Based Theory of Language Acquisition* (2003) [18], children are not born with language structures in their brains. Rather, language structures emerge through using language. Relying on general cognitive skills and learning mechanisms, children identify and use rules from the language around them in order to establish their own mental set of those rules. In this way, children gradually discover the language of the communities they belong to and "construct" linguistic knowledge. Seen from this perspective, language is an external object which becomes internalised (**internalisation**). According to Ibbotson and Tomasello [19], much research now shows that young children start by learning simple grammatical patterns such as the following:

Where's X?
X is here.
X is gone.
Mummy's X-ing it.
Mummy X-ed it.
Let's X it.

Later, children combine these simple patterns into more complex ones, e.g.

Where's the X that Mummy X-ed?

The researchers claim that if children were following innately given rules, they would not make errors inconsistently. However, many children make errors such as *Why he can't come?* while at the same time correctly saying *What does he want?* Ibbotson and Tomasello conclude that "children are not born with a universal, dedicated tool for learning grammar" but "a set of general-purpose tools, such as categorisation, the reading of communicative intentions and analogy making, with which they build grammatical categories and rules from the language they hear around them" [19].

Internalisation: Converting what is perceived externally into a feature of one's interior (cognitive) landscape. For example, the rules of grammar of one's native language are internalised as linguistic knowledge.

The "connectionist" approach is slightly different; it suggests that language is learned through exposure to patterns and regularities in linguistic input which become strengthened by repeated activation. There is no rule learning as such: learning is simply the result of strengthening connections between neurons in the brain. Researchers working within this framework often devise computer models (neural networks) to test their hypotheses. For example, Rumelhart and McClelland [20] simulated learning regular versus irregular past tense forms in English using a computer that made generalisations based on stored examples. This reproduced fairly accurately the three phases that children go through to acquire the past tense:

1. Irregular past tense forms (e.g. *went, fell*) are first produced correctly through rote-learning.
2. The regular past tense ending -ed is then overgeneralised to irregular verbs (e.g. *goed, falled*)
3. Irregular forms are supplied correctly again [20].

Rumelhart and McClelland's simple learning model has since been applied not only to morphology but also to phonology, prosody, syntax and the lexicon. What distinguishes the connectionist approach is that it is neurally inspired and data driven and regards language as the outcome of pattern recognition and statistical learning rather than as a pre-specified system.

In stark contrast to the constructionists and connectionists, the nativist Noam Chomsky argues that human language is far too complex to be learned from the input data and that we must therefore have an "innate predisposition" to expect language to be structured the way it is. He proposes that there are certain universal "principles" that underlie all languages, and these principles are hardwired into the human brain so that they do not have to be discovered. An example of one is the principle of structure-dependency (Chomsky 1965) [21], which holds that words and phrases are grouped in a hierarchical organisation (as in phrase structure grammar). Another, the "Move" principle (Chomsky 1981) [22] holds that whenever elements of the sentence are moved to form questions, passives, etc., this movement takes into consideration this hierarchical organisation, as the following example demonstrates:

> **The cat he found** is black and white.
> Is **the cat he found** black and white?
> * **The cat** is **he found** black and white?

In the above sentences, "the cat he found" is organised in a phrase that cannot be split when the verb is moved to form a question. These principles, collectively known as **Universal Grammar (UG)**, supposedly provide a uniform initial state and foundation for the learning of any language in the world.

> **Universal Grammar (UG)**: Theory of an innate biological component of language proposed by Noam Chomsky. Set of principles that underlie all languages that are hardwired into the human brain.

The theory continues: on being exposed to a particular language, children use an innate "**Language Acquisition Device**" (**LAD**) to set "parameters" of UG (Chomsky 1981) [22]. For example, the "head" parameter relates to the way phrases are organised. Each phrase has a central element, called a head, and a complement. In the verb phrase "love sushi", the verb "love" is the head and "sushi" the complement. In the prepositional phrase "in Tokyo", the preposition "in" is the head and "Tokyo" the complement. English is a head-first language because the head of the phrase always precedes its complement ("in Tokyo", not "Tokyo in"). Japanese, on the other hand, is a head-last language because heads follow their complements. Thus, in Japanese, we have:

私は	寿司が	大好き	
I	sushi	love	(= I love sushi)

私は	東京	に	住んでいる	
I	Tokyo	in	live	(= I live in Tokyo)

According to Chomsky, children endowed with UG do not have to figure out that language is structured in phrases. The only task left is to discover which parameter settings apply in the language they are learning, a considerably simpler task than if they had to start learning a language from scratch. There is evidence from first language acquisition research that English-speaking children have set the "head" parameter as early as the two-word stage, as shown by the following utterances of one- to two-year-olds, taken from Radford (1990) [23]:

Head	Complement	
Open	*box*	(= Open the box)
Get	*toys*	(= Get my toys)
In	*there*	(= Put it in there)
Out	*cot*	(= I want to get out of the cot)

Language Acquisition Device (LAD): Innate mental capacity proposed by Noam Chomsky for acquiring language.

From an evolutionary perspective, there is a major sticking point with Chomsky's theory: if language is biologically determined, as he claims it is, the emergence of such complexity would not have been possible within such a short time span (presumably 100–200 thousand years) [24]. Chomsky's Minimalist Program from the 1990s tried to answer this problem by suggesting that the only "given" by UG is the computational system that grammar is subject to. In its original version, Minimalism reduced syntax to a single computational operation called "**Merge**". Merge takes two syntactic objects (words or word sets) and forms a new set with them:

$$\text{drink}_V + \text{water}_N \rightarrow V \{\text{drink}_V \text{ water}_N\}$$

The operation is **recursive** in that the output becomes the new input:

$$V \{\text{drink}_V \text{ water}_N\} + P \{\text{from}_P \text{ a}_D \text{ stream}_N\} \rightarrow$$
$$V \{\text{drink}_V \text{ water}_N \text{ from}_P \text{ a}_D \text{ stream}_N\}$$

As Merge operates successively, it builds the syntactic structure of a sentence from the bottom (lexical items) to the top (complete sentence). In this view, speech works "bottom up", i.e. it occurs in a more piecemeal fashion, word by word, without advance knowledge of upcoming structure. This seems to be totally at odds with Chomsky's earlier work on principles and parameters, stressing structure-dependency. Chomsky proposes that Merge is a binary operation since it combines two elements at a time. Binary branching may be computationally efficient but whether it is cognitively plausible remains an open issue. Moreover, this binary operation is claimed to be the result of a sudden change in the human genome, and the only new function that was necessary for language to evolve, despite there being empirical evidence that recursion is neither unique to language nor to humans.

Minimalist Program: Model of human language proposed by Noam Chomsky that assumes the core feature of language is a human-specific computational system (known as Merge) that allows us to create an infinite number of novel sentences.

Merge: A structure-building operation in the Minimalist Program that allows us to tag words onto words recursively to create an infinite number of expressions.

Recursion: The repeated coupling of elements (e.g. words) to create larger elements (e.g. phrases). Thought by some to be responsible for producing a boundless system of communication in humans.

Chomsky's work has been variously interpreted and critiqued over the years, even to the point of being labelled "an unscientific revolution". According to Mitchell and Myles [25], one of the main criticisms of the generative framework is that it is syntactocentric, dealing only with "the sentence and its internal structure, rather than any larger unit of language". In addition, it studies language "in a vacuum, as a mental object, rather than a social or psychological one" [25]. While Universal Grammar is doubtless a sophisticated tool for decomposing language in order to make testable hypotheses, it remains limited in that it is primarily a theory of what language is like. It is not a theory of how linguistic structure is derived, nor how it develops. It is now generally recognised that theories of language evolution also need to consider the learner's mind as a processor and learner of information, in addition to the linguistic information it encapsulates.

Thus, to understand the origins and development of language, we need to study the cognitive capacities underlying the ability to use and acquire language. There is, however, one more crucial dimension to language. It is not simply a psychological object that belongs to the individual but a social one that needs to be learnt. And for learning to take place, language needs to be externalised.

8.5   Externalisation and Cultural Transmission

Before language can be used for interaction, it needs to be **externalised** by the sensorimotor system, which is dependent on exposure to the system and imitation. This presents a paradox for language evolution. How could humans' first words have been externalised if humans had no words to copy? The solution to this paradox can be explained if we assume that humans' first deliberate words were onomatopoeic, i.e. they imitated the sounds of the objects and actions they were referring to.

Externalisation: Expressing outwardly what is originally internal. For example, linguistic knowledge can be externalised as speech.

Today's children are of course exposed to full language from birth, but how does a young child make sense of it all? As mentioned at the beginning of the chapter, children have a predisposition to find patterns in the environment, which makes them active, self-motivated learners. Of course, no parent gives their baby explicit grammar lessons, but most parents do unwittingly have a range of tricks up their sleeves to help matters. **Child-directed speech**, formerly referred to as motherese, is a special way of talking to your child. It is characterised by a slower pace, exaggerated pitch modifications and the use of special vocabulary (e.g. *beddy-byes*), diminutives (e.g. *doggie*) and mimicking words (e.g. *woof woof*). The function of child-directed speech is varied: firstly, it serves to gain and hold an infant's attention; secondly, it is thought to play an important role in the emotional bonding between parents and child, and lastly, there is evidence that it helps in language learning by emphasising the structure and meaning of language.

Child-directed speech: Special way of talking to an infant, characterised by a slower pace, shorter utterances, exaggerated pitch modifications and special vocabulary.

Japanese researchers Mutsumi Imai and Sotaro Kita [26] consider **sound symbolism** (a form of iconicity where a word's sound evokes its referent) to be an important design feature of language that helps children to acquire their mother tongue. The researchers note that young infants are more sensitive to sound symbolism than older children and this helps them "gain the insight that speech sounds refer to entities in the world" [26]. They also entertain the idea that language acquisition in children mirrors how language evolved in our distant ancestors. They suggest that the sound symbolism in motherese is the remnant of a protolanguage that was predominantly sound symbolic. They claim that humans would have first mimicked the external world. This use of motivated form would have then helped our ancestors to build a shared lexicon that was easily understood by members of the community [26]. As the lexicon grew, it would have been difficult for language to maintain sound symbolism for all words, hence the switch to arbitrariness and a greater reliance on memory and vocal learning. Thus, sound symbolism may have set the foundation for humans to use speech sounds to systematically refer to concepts.

Sound symbolism: Perceptual similarity between the sound of a sign and its referent. The sound of a sign may imitate the sound of the object it refers to (onomatopoeia) or some other sensory property of the object, such as size, shape or movement (e.g. the /k/ and /t/ sounds in "cut" imitate sharpness.

A final insight into the nature of language and language acquisition can be gained from the deaf community. Firstly, the fact that communication can work perfectly through sign language (which uses the visual-gestural modality instead of the vocal-auditory) shows that the precise channel through which language is perceived and externalised is of secondary importance. What is crucial is some form of cross-modal processing. Secondly, it has been observed time and time again that when deaf children are denied exposure to sign language users, they develop a system of gestures that are understood only by the child and its immediate caregivers. In addition, we know that these children, once they are exposed to a standardised sign language, will often adopt the new language and lose their original one. These two facts are a clear indication that language is a highly innate characteristic of the human species, but one that is dependent on exposure and social interaction. Without a genetic base for processing and learning and a social environment from which to learn, language could not have evolved.

8.6 Chapter Summary

Language acquisition studies suggest that children are born with a subtle mix of innate cognitive mechanisms, including categorical perception, imitation and general learning skills, that provide a biological framework for language to develop in the individual. However, language learning cannot take place without social interaction and exposure to the language being learnt and is therefore dependent as much on the external environment as on internal constraints. This creates a paradox for the evolution of language: how could humans' first words have been learnt if there were no words to copy? The solution to this paradox can be explained if we assume that humans' first deliberate words were onomatopoeic. Using sounds to imitate the noises of objects may have been the key to using speech sounds to systematically refer to concepts.

In these last two chapters, we have examined the cognitive processes involved in speech perception, speech production and language acquisition in humans. As research progresses, we are discovering more and more that the building blocks of these processes can be found in other animals and are therefore likely to have been inherited from our primate ancestors. In the next chapter, we will compare human language with other animal communication systems to find out which features of our language are shared with other species and which are truly unique to humans.

References

1. Hartley, A. F. (1982). *Linguistics for language learners*. The Macmillan Press.
2. Mehler, J., Jusczyk, P., Lambertz, G., Halsted, N., Bertoncini, J., & Amiel-Tison, C. (1988). A precursor of language acquisition in young infants. *Cognition, 29*(2), 143–178. https://doi.org/10.1016/0010-0277(88)90035-2
3. Mampe, B., Friederici, A., Christophe, A., & Wermke, K. (2009). Newborns' cry melody is shaped by their native language. *Current Biology, 19*(23), 1994–1947. https://doi.org/10.1016/j.cub.2009.09.064
4. Friederici, A. D., Friedrich, M., & Weber, C. (2002). Neural manifestation of cognitive and precognitive mismatch detection in early infancy. *Neuroreport, 13*(10), 1251–1254. https://doi.org/10.1097/00001756-200207190-00006
5. Friedrich, M., Weber, C., & Friederici, A. D. (2004). Electro-physiological evidence for delayed mismatch response in infants at-risk for specific language impairment. *Psychophysiology, 41*(5), 772–782. https://doi.org/10.1111/j.1469-8986.2004.00202.x
6. Friederici, A. D., Friedrich, M., & Christophe, A. (2007). Brain responses in 4-month-old infants are already language specific. *Current Biology, 17*(14), 1208–1211. https://doi.org/10.1016/j.cub.2007.06.011
7. Pannekamp, A., Weber, C., & Friederici, A. D. (2006). Prosodic processing at the sentence level in infants. *Neuroreport, 17*(6), 675–678. https://doi.org/10.1097/00001756-200604240-00024
8. Tallerman, M. (2007). Did our ancestors speak a holistic protolanguage? *Lingua, 117*, 579–604. https://doi.org/10.1016/j.lingua.2005.05.004
9. Kamhi, A. G. (1986). The elusive first word: The importance of the naming insight for the development of referential speech. *Journal of Child Language, 13*(1), 155–161. https://doi.org/10.1017/S0305000900000362
10. Friederici, A. D. (2011). *Neuropsychologische Grundlagen der Sprachentwicklung*. Max-Planck-Institut für Kognitions- und Neurowissenschaften.
11. Bergelson, E., & Aslin, R. (2017). Nature and origins of the lexicon in 6-mo-olds. *PNAS, 114*(49), 12916–12921. https://doi.org/10.1073/pnas.1712966114
12. Jackendoff, R. (2002). *Foundations of language: Brain, meaning, grammar, evolution*. Oxford University Press.
13. Goodall (1986). *The chimpanzees of Gombe: Patterns of behaviour*. Belknap.
14. Arbib, M. A., Liebal, K., & Pika, S. (2008). Primate vocalization, gesture, and the evolution of human language. *Current Anthropology, 49*(6). https://doi.org/10.1086/593015
15. Hirsh-Pasek, K., & Golinkoff, R. (1996). *The origins of grammar: Evidence from early language comprehension*. MIT Press.
16. Yuan, S., & Fisher, C. (2009). "Really? She blicked the baby?" Two-year-olds learn combinatorial facts about verbs by listening. *Psychological Science, 20*(5), 619–626. https://doi.org/10.1111/j.1467-9280.2009.02341.x
17. Brown, R. (1973). *A first language: The early stages*. George Allen & Unwin Ltd.
18. Tomasello, M. (2003). *Constructing a language: A usage-based theory of language acquisition*. Harvard University Press.
19. Ibbotson, P., & Tomasello, R. (2016). Evidence rebuts Chomsky's Theory of language learning. *Scientific American*. https://www.scientificamerican.com/article/evidence-rebuts-chomsky-s-theory-of-language-learning/

20. Rumelhart, D., & McClelland, J. (1986). On learning the past tense of English verbs. In J. McClelland & D. Rumelhart (Eds.), *Parallel distributed processing: Explorations in the microstructure of cognition* (Vol. 2: Psychological and biological models, pp. 216–71). MIT Press.
21. Chomsky, N. (1965). *Aspects of the theory of syntax*. MIT Press.
22. Chomsky, N. (1981). *Lectures on government and binding*. Foris.
23. Radford, A. (1990). *Syntactic theory and the acquisition of English syntax*. Basil Blackwell.
24. Fujita, H., & Fujita, K. (2021). Human language evolution: A view from theoretical linguistics on how syntax and the lexicon first came into being. *Primates*. https://doi.org/10.1007/s10329-021-00891-013
25. Mitchell, R., & Myles, F. (2004). *Second language learning theories*. Arnold.
26. Imai, M., & Kita, S. (2014). The sound symbolism bootstrapping hypothesis for language acquisition and language evolution. *Philosophical Transactions of the Royal Society B, 369*, 20130298. https://doi.org/10.1098/rstb.2013.0298

Chapter 9
How Unique is Human Language?

This chapter takes the internalist (cognitive) perspective and concentrates on the biological foundations of language. A fundamental question in the study of language origins is which aspects of the language capacity are inherited from other species and which evolved later and only in humans. According to the saltationists, the only uniquely human component of the language faculty is the computational mechanism of recursion that enables the repeated coupling of words to create phrases and sentences. This capacity, they claim, appeared as the result of a sudden change in the human genome (the precise mechanism not being specified) and should be the primary interest of linguistics, which, after all, seeks to understand the fundamental nature of human language. In 2002, Hauser, Chomsky and Fitch wrote a seminal article [1] that attempted to integrate Chomsky's nativist and discontinuity perspective (focusing on the uniqueness of language) with the biologist's adaptationist and comparative perspective (which emphasises continuity with other species). According to the article, however, recursion remains the key component of the human language faculty, constituting the "core" of language, or "**Narrow Language Faculty**" (**FLN**); all other features of language, including a wide range of cognitive and perceptual mechanisms shared with other species, belong to the "**Broad Language Faculty**" (**FLB**) and are of secondary importance, simply interacting with the recursive core to facilitate language use.

> **Narrow Language Faculty (FLN)**: Proposed "core" of language which is unique to humans and specific to language.

> **Broad Language Faculty (FLB)**: Whole set of mechanisms involved in acquiring, perceiving and producing language.

J. Dornbierer-Stuart, *The Origins of Language*,
https://doi.org/10.1007/978-3-031-54938-0_9

From the opposite camp, Pinker and Jackendoff [2] believe there is considerably more of language that is special to humans. For them, language is a system of inherited traits which have been reshaped in humans by natural selection and which have co-evolved gradually, via protolanguage, for the purpose of communication. This chapter takes the broader biological approach and assumes language is multicomponential, complex and slow-developing. We will therefore be interested in the whole set of mechanisms, including the vocal/motor, auditory/perceptual and central neural, that are involved in the processing and acquisition of language. We will start the chapter by looking at Hockett's comparative analysis of human language and other animal communication systems. We will then enlarge upon the comparative approach by comparing the cognitive capacities across living species today, with a view to understanding the evolutionary path to human language. As we shall see, many of the mechanisms involved in human speech are in fact built upon ancient capacities inherited from other species but which have been subsequently modified in humans.

9.1   Hockett's Behaviouristic Approach

Given language is the primary means of communication among humans, it makes sense to start off by comparing human language with other animal communication systems. In the 1950s and 60s, the American linguist and anthropologist Charles Hockett developed a set of design features of language to do just that. He wanted to identify points of connection but also determine which features were unique to human language. His approach was zoological, concentrating on observable behaviours, with a distinct naturalistic sentiment: he wanted to find "man's place in nature". He researched six systems of communication, including bee dancing, birdsong, gibbon calls and human language, and came up with a list of "key properties of language". Some, he claimed, were shared by humans and many other animals (e.g. use of the vocal-auditory channel), some were shared only by humans and other primates (e.g. semanticity, arbitrariness and discreteness) and some were unique to human language (e.g. duality of patterning, cultural transmission, productivity and displacement) [3]. This seems to suggest a natural evolutionary progression towards full language (Fig. 9.1).

Let us start with (and revisit) the feature of duality of patterning (the combining of a small number of meaningless elements to produce a large number of meaningful elements), which Hockett claimed was exclusive to humans. Generally, it was believed that animals that use vocal signals have a stock of basic sounds, and the number of possible messages is restricted to the number of sounds. This assumes there is no combining of sounds to create new messages. Hockett found no clear evidence of duality in any of the non-human systems he studied, but he was not quite sure about birdsong. Of course, there is combining of sounds in birdsong: individual tweets are combined to form different melodies. However, neither the tweets nor the

	Western Meadowlark Song	Gibbon Calls	Human Language
Duality of Patterning	?	X	
Cultural Transmission	?	?	
Productivity	?	X	
Displacement	?	X	
Discreteness	?		
Arbitrariness	?		
Semanticity	?		
Vocal-auditory Channel			

Fig. 9.1 Comparison of three communication systems (adapted from Hockett 1960) [3]

songs have referential meaning, i.e. they are not used to refer to things in the world—they are just used holistically to impress, defend or send warnings. This feature of combining sounds without referential meaning is termed **bare phonology**.

> **Bare phonology:** Patterning of sounds in holistic messages or tunes that refer to whole ideas or situations and cannot be broken down into smaller meaningful parts.

9.2 Shared Evolutionary Roots

The problem with Hockett's approach was that he was simply comparing the codes of different communication systems instead of looking at the "abilities" required by the code users. Modern research takes the evolutionary approach and tries to understand the cognitive abilities and biological adaptations that were necessary for the system to evolve. There is already considerable evidence that many of the mechanisms comprising the language faculty have very deep evolutionary roots, being present long before humans emerged. This ties in with the concept of **continuity**, which holds that language has evolved from precursors in the animal world. It can be argued that human speech builds on 200-million-year-old brainstem circuitry shared by most vertebrates that enables instinctive emotional vocalisations. A significant adaptation would have been the development of mirror neurons in the motor cortex of primates some 45 million years ago. This enabled the imitation of actions and tool-use learning in apes, considered to be a precursor to vocal imitation and language learning in humans [4]. Such **"homologous traits"** (traits handed down from other species) did not evolve specifically for human language but are now an integral part of the language faculty.

> **Continuity**: View that language has evolved from precursors in the animal world.

> **Homologous trait**: Feature shared with another species through a common ancestor.

Researchers Zuidema and de Boer [5] suggest that many of the biological prerequisites for duality are inherited from other animals. One prerequisite would have been the capacity for combinatorial structure (see Sect. 2.1), which itself requires multiple components to be put in place. First, there needs to be "a repertoire of basic elements shared by sender and receiver"; second, "a mechanism to combine those elements into larger combinations in the sender" (**synthesis**), and third, "the mechanisms to break down combinations into their component parts in the receiver" (**analysis**) [5].

> **Synthesis**: Putting pieces together as a whole.

> **Analysis**: Breaking down a whole into its component parts.

We will now look at some of the evidence suggesting that the roots of combinatorial structure have been inherited from the animal world, starting with auditory perception.

9.2.1 Auditory Perception

As for Zuidema and de Boer's first component of combinatorial structure (a shared repertoire of sounds), Hockett recognised that both humans and apes make use of **discreteness** (the use of a limited set of discrete sounds along the continuum of possible sounds). In 1975, Kuhl and Miller famously demonstrated that chinchillas (a type of rodent) perceive human speech sounds categorically in much the same way as we do. This means, for example, sounds along the voiced b to voiceless p continuum are perceived as either /b/ or /p/—there is no middle ground [6]. Parallel research on categorical perception in apes has shown that they can perceive and discriminate many sounds of human language that they cannot produce. This implies that the hearing prerequisites for speech perception were largely in place before humans split from chimps. According to the American evolutionary biologist and cognitive scientist W. Tecumseh Fitch, research on a wide variety of birds and mammals suggests that "the perceptual apparatus underlying speech is simply the human version of a general voice-perception system shared with other species" [7].

Discreteness: A feature of language that uses a limited set of sounds far enough apart on the continuum of possible sounds so that each sound remains distinctive.

According to Pinker and Jackendoff [2], however, speech perception in humans goes far beyond the ability to perceive and discriminate speech sounds. Humans do not just make "one-bit discriminations between pairs of phonemes" but can distinguish individual words and extract meaning from a continuous stream of speech [2]. Besides, there is considerable evidence that speech sounds are processed separately from other sounds. Neuroimaging suggests that speech sounds and non-speech vocal sounds (cries, laughs, sighs) are served by different areas of the brain. This implies that the speech perception system in humans has evolved independently since our split from chimps.

9.2.2 Combinatoriality

As for Zuidema and de Boer's second component of combinatorial structure (a mechanism to combine elements), Hockett recognised that birds use patterning (combinatoriality) in their songs, although he found no evidence of this feature in our closer relatives, the apes. It is now known that the patterning in birdsong is surprisingly complex. According to Beecher [8], song sparrows typically have nine different songs; each of these songs is made up of five or six distinct elements, and the order is important (like syntax). The different songs are used to escalate and de-escalate disputes following a set of conventions based on which songs are shared with enemies (used for escalation) and which are not (used for de-escalation). However, no bird rearranges its tweets into different songs that signify different things [8].

Although Hockett found no evidence of combinatoriality in gibbon calls, it has since been detected in the calls of chimpanzees. According to Leroux and Townsend [9], chimps combine roughly 15 calls into well over 80 combinations, but, again, these calls are used to signal such things as identity and status rather than convey novel messages [9]. This has led to the suggestion that the same thing happened in the evolution of human language, that combinatorial speech was originally holistic, used for display purposes, before it became semantically productive. Zuidema and de Boer suggest that semantically productive combinatorial structure (combinatorial phonology) (see Sect. 4.5) can emerge from a holistic communication system when there is repeated interaction between individuals and a growing number of concepts that need to be expressed [5], but this is questionable if we consider that speech and holistic vocalisations are served by different areas of the brain.

9.2.3 Semanticity

Zuidema and de Boer's third component of combinatorial structure (a mechanism to break down combinations into their component parts) would have required the ability to assign meaning to sounds. Hockett recognised that gibbon calls as well as human language exhibit both **semanticity** (the use of signals to represent events, ideas, actions and objects in the real world) and arbitrariness (the absence of any natural connection between a sign and its meaning). It is now well known that the various calls of vervet monkeys (native to Africa) refer to different predators. According to Seyfarth et al. [10], a chattering sound is used to warn other group members against snakes, and a chirping sound warns against lions or leopards [10]. Since the calls are used to refer to certain creatures in the real world, they can be ascribed semanticity. Likewise, since the acoustic structure of the calls does not resemble the predator in any way, the calls can be credited with arbitrariness, rather like words. However, the vervet monkeys' alarm system consists of a very limited number of calls produced in a narrow range of contexts. The fact that they are fairly rigidly coupled with danger suggests they are purely symptomatic and thus closer to the more primitive stimulus–response reaction rather than the higher cognitive capacities associated with human language.

> **Semanticity**: Ability of language to use signals to represent events, ideas, actions and objects in the real world.

A much clearer case of semanticity and arbitrariness in non-human primates comes from studying the visual rather than vocal communication of chimpanzees [11]. Experiments have shown that chimps are able to complete tasks requiring them to respond appropriately to a set of arbitrary symbols that represent words (lexigrams), e.g. "give Sarah apple". Kanzi the bonobo (now aged 43), raised by primatologist Sue Savage-Rumbaugh, has been trained to recognise both lexigrams and spoken words and can follow simple instructions such as "Put the apple in the fridge". However, many believe that this behaviour could be the result of a sophisticated imitation ability rather than some form of linguistic processing. While chimps may understand individual symbols or words, they cannot grasp how words are assembled to form a complete idea, nor can they create novel sentences. And for language-like behaviour to emerge in these animals requires "unusually rich human-like developmental conditions" [12].

9.3 Convergent Evolution

So far, we have seen that many of the building blocks for human language are found in other primates, suggesting shared inheritance from a common ancestor. It is also possible that a trait recognised as relevant to language evolved convergently (independently) in humans and other species. **Convergent evolution** creates **"analogous traits"**, i.e. features that have similar functions among diverged species but that were not present in the common ancestor of the two. For example, the syrinx, the unique vocal organ found in birds, is thought to have evolved to supplement sound production in the longer neck of birds, followed by loss of the larynx. In humans, the permanent descent of the larynx (see Sect. 4.2) may have served a similar purpose for human language.

> **Convergent evolution**: Independent development of similar features in species with different ancestral origins.

> **Analogous trait**: Similar feature that evolved independently among genetically distant species.

Other features of language that evolved convergently in birds and humans include combinatoriality and vocal learning. Although these features evolved at different times in vertebrate evolution, they are interesting for evolutionary linguists because similar neural mechanisms are involved. Birdsong can therefore give us vital clues to how our language developed, despite the fact that songbirds evolved at least 50 million years before humans. We will now look more closely at vocal learning in birds to see what insights can be gained for the origins and development of this feature in humans.

9.3.1 Vocal Learning

Most songbirds are not born with their songs but have to learn them by listening to and imitating other birds, usually their fathers or other adult males in their social environment. There are several striking similarities between **vocal learning** in birds and humans, especially with regard to how sensory experience is internalised and used to shape vocal outputs. Auditory responses in avian vocal motor neurons suggest that birds perceive their song as a series of articulatory gestures, in line with the "motor theory of speech perception" (see Sect. 7.4). Another striking similarity relates to a "babbling" phase (see Sect. 8.1). In both birds and babies, this involves the production of repetitive, nonsensical vocalisations that serve as a precursor to, and possibly a prerequisite for, more structured and mature vocal communication.

Vocal learning: Learning of language or birdsong through social interaction using innate mechanisms such as imitation, memory and general learning strategies.

It has been argued by Deacon [13] that the relaxation of natural and sexual selection on birdsong production (e.g. due to ecological stability) was responsible for its complexification. With a reduction in the innate biases controlling song production, attentional factors would have begun to influence vocal learning, resulting in birdsong becoming more variable and complex. The fact that human infant babbling occurs in contexts of low emotional arousal might also be allowing more brain systems to participate in vocal learning [13]. However, this idea remains speculation and is not universally accepted.

Research is now showing that humans may have actually "inherited" some of the prerequisites for vocal learning from apes. Ayumu is a chimpanzee currently living at the Primate Research Institute of Kyoto University that has been taught to recognise at least a dozen numbers and put them in ascending order. Much more impressive than this is his performance in the "working memory test". In this test, the numbers 1 to 9 are randomly scattered across a touch screen and momentarily displayed. The chimp quickly studies the layout of the numbers and, as soon as they are blanked out, has to touch the position of each number, in ascending order. Ayumu's skill in this task is far superior to that of comparably trained university students, suggesting that chimpanzees have a better working memory than humans [14].

9.3.2   Cultural Transmission

A final point of comparison between birdsong and language relates to the iterative nature of learning, i.e. the fact that the output of one generation becomes the input of the next generation. This brings us to another one of Hockett's design features, cultural transmission, which he defined as the passing on of vocal behaviour from generation to generation by learning and teaching. Hockett found no evidence of this feature in non-human communication, but research has since shown that birds raised in isolation sing differently, emphasising the role of tutors in learning birdsong. Marler and Sherman [15] found that learning affects all levels of birdsong (including song repertoire size, song duration, the number of notes per song and the duration of notes and inter-note intervals), but has most influence "at the level of the fine structure of the notes and syllables from which the songs are constructed" [15]. This is significant because it highlights the importance of flexibility and adaptability in communication systems. Both birdsong and human language are dynamic and can be influenced by learning, allowing for variation and modification over time.

It should be noted that vocal behaviour is only one of many forms of culture that can be transmitted socially. Recent research has shown that apes, too, transmit culture through social learning. For example, Gruber et al. [16] demonstrated that two different Ugandan chimpanzee societies with the same tools at their disposal adopted two different approaches to get honey in the cavity of a fallen log. Individuals of the Sonso community of Budongo Forest manufactured leaf sponges to use as probes while individuals of the Kanyawara community of Kibale National Park manufactured sticks to extract the honey. The researchers concluded that the tool choices of each group depended on pre-existing cultural differences in tool-use behaviour [16]. This implies that the behaviour emerged through social learning, thus the prerequisites for vocal learning in humans could have been inherited from apes.

Social learning can be demonstrated in the lab. Experiments have shown that both humans and other primates create "chains of learning" from one generation to the next, whereby changes occur as each individual interprets and modifies the information being passed on (as occurs with language change). Going a step further, it can be demonstrated how structure emerges from scratch through the process of generalisation. For instance, when a human is tasked with learning a vocabulary of nonsense words and meanings, and the learnt output then becomes the input for the next participant, it is usual that the new participant works with only a subset of the vocabulary and generalises patterns to unseen items, creating structure. Similarly, baboons can be tasked with learning simple random patterns on a grid, and their test output then becomes the input for another baboon. Eventually, the patterns are no longer random but exhibit structure [12]. This is strong evidence that language is an external object passed on culturally, independent of biological evolution, yet it owes its existence to innate learning abilities inherited from the animal world. Language exists in the interaction between these two domains.

Returning to birds, there are of course also major differences between birdsong and language. As we saw in Sect. 9.1, the elaborate and structured songs of birds do not refer to things in the world—they are just used holistically to attract mates or establish territory. Parrots, on the other hand, do not learn highly structured songs to mark territory but have another exceptional vocal skill. They are known for their vocal mimicry, which involves imitating a wide range of sounds, including human speech, other bird calls and environmental noises. Their mimicking behaviour is thought to play a role in social interactions, establishing bonds, and even problem-solving, all important aspects of human language.

It could be argued that in nature, parrots, like most other species, have a disadvantage when it comes to learning language because they have a limited time to learn—they have to get on with the business of survival. But given the chance, parrots can learn basic human language. Cognitive biologist Dr. Irene Pepperberg worked with Alex, an African grey parrot, on a thirty-year project that produced some surprising results. Alex was able to name objects, colours, materials and shapes, identify numbers and letters and answer questions such as "How many?" "What's different (colour or size)?" "What's bigger?" "What's smaller?" "What material?" etc. This not only suggests advanced working memory but even the ability to form hierarchical sets. Moreover, Alex was able to correctly answer questions designed for six-year-old humans [17]. Alex's behaviour might be seen as a primitive form of linguistic processing, although his skill faltered when it came to manipulating signs to produce new sentences. It seems only humans are able to combine all the features necessary for language.

9.4 Divergent Evolution

Clearly, there are also features underlying language that evolved later and only in humans. **Divergent evolution** is the term used for the development of differences within a species that can lead to speciation. The concept of **discontinuity** (whereby a biological capacity evolves from scratch in a species) can be applied to language in that it comprises a set of novel features not found in our closest relatives. While some of these unique features, or **derived traits**, relate to vocal anatomy (e.g. lowering of the larynx, development of the tongue), many of the principal changes required for the evolution of speech since our divergence from chimpanzees are neural.

Divergent evolution: Development of differences within a species that can lead to speciation.

Discontinuity: Principle that a biological capacity in a species can evolve from scratch in that species.

Derived trait: Feature evolving uniquely in a species, e.g. the large brain in humans.

Paleoneurological studies, which involve making endocasts (real or virtual) of the insides of archaic human skulls and comparing these to the morphology of modern human brains, suggest that our species, Homo sapiens, has a unique organisation of the parietal areas of the brain. These areas are known to be engaged in higher cognitive functions, allowing us, for example, to understand and memorise a vast number of new words and combine them creatively in novel messages that will be instantly understood by others. Unfortunately, the downside of a highly evolved brain is that it seems to increase vulnerability to dysfunctional brain architecture and neurodegenerative processes leading to disorders such as autism, schizophrenia and dementia.

9.4.1 Productivity and Displacement

Hockett rightly regarded language as being distinct from all other known animal forms of communication because of its creativeness. Humans are constantly manipulating language to describe new objects and situations, coming up with new expressions and novel utterances. This open-endedness, or **productivity**, as Hockett called it, means there is no limit to the number of possible utterances in any human language. Most of the utterances you produce and hear every day have most likely never been produced by anybody else.

Productivity: The ability of language to make novel utterances.

Hockett nevertheless confused matters by showing that the feature of productivity is present in the behaviour of the honeybee. The dance it performs when it returns to the hive conveys precise information about the location and amount of nectar it has discovered. What's more, this communication system shares another one of Hockett's defining features of human language, that of displacement (see Sect. 3.1), since it is communicating about something that is remote in space and time. However, in evolutionary terms, we are far too distant from bees to expect human language to have evolved directly from bee dancing. It actually makes much more sense to look at comparable features in non-human primates.

A much more relevant example of displacement can be found in chimpanzees in the sphere of **cognition**. Researchers Mathias Osvath and Helena Osvath [18] demonstrated future planning in chimpanzees, who were shown to sacrifice a small food reward to obtain a tool they could use to retrieve a larger food reward. According to Osvath and Osvath, planning for future needs relies heavily on two overarching capacities: self-control, defined as "the suppression of immediate drives in favour of delayed rewards", and mental time travel, or the ability to pre-experience in the mind a future event [18]. The ability to plan ahead could well have been the necessary ingredient for using signals to refer to something beyond what is immediately present.

Cognition: All functions and processes of the mind, including perception, attention, thought, imagination, intelligence, memory, reasoning, problem-solving, decision-making and language.

We will explore the roots of human cognition further in Sects. 9.6 and 9.7, but for now, we will turn to a set of traits related to vocal production that have clearly emerged since our divergence from apes. Up to now, no non-human primate has ever been able to produce more than three human words.

9.4.2   Vocal Production

A feature that developed in humans after our split from chimpanzees (although present by convergent evolution in birds) is our superior vocal control. Although chimpanzees can boast some of the most sophisticated cognitive abilities reported among animals, chimpanzee **vocal production** does not exceed that of many other primates or mammals, with a small repertoire of 30-odd innate vocalisations [19]. Evidence that vocal production has been adapted in humans for speech comes from studying the vocal tract and the brain. According to vocal tract specialist Philip Lieberman, the descended larynx is one of a suite of vocal tract modifications relevant to speech, although it has been proposed that the original purpose of the descended larynx was to lower pitch in order to convey a sense of greater physical size to deter rivals, rather than specifically to enable language. Other modifications include changes in the shape of the tongue, that would have increased flexibility for artic-ulation, and an enlarged pharyngeal cavity, that would have expanded the space of discriminable speech sounds [20].

Vocal production: Ability to articulate speech sounds.

As regards the articulation of speech sounds, primitive vowels and consonants are found in other animals. Chimps can produce vowel-like sounds, albeit not very distinct ones. Humans have the advantage of a chunky muscular tongue, which allows them to produce the three extreme vowels [i], [a] and [u]. These key vowel sounds alone make it possible to distinguish many words. As for consonants, primate lip-smacks use in essence the same mechanism as humans do for [p], [b] and [m], but again, humans are better endowed to produce a far greater range of consonants [21].

According to Fitch [7], it is now thought by many that the importance of vocal anatomy has been overemphasised and that perhaps the most significant changes required for the evolution of speech since our divergence from chimpanzees were neural. While brain expansion is commonly believed to have started about 2.5 million years ago (around the start of the Pleistocene), scientists have found that the brains of early Homo retained an ape-like frontal lobe for much longer than originally estimated—until about 1.5 million years ago (see Sect. 4.4). Development of the frontal lobe (the "action" lobe) has been linked specifically to the development of speech control in humans. Some have suggested greater frontal lobe development occurred at the expense of posterior ("sensory") areas of the brain. This would have supported the ability to ignore ongoing sensory-motor processing in order to simulate non-present situations, i.e. it would have enabled the capacity for displacement.

Possibly, though, the most critical derived trait for human language is the increased **cerebral connectivity** that enables us to establish neural connections between all the brain regions required for language. Without an overall control system, speech cannot work. Cortical folding is crucial for the brain circuitry and its functional organisation. While monkeys and apes have many of the pre-linguistic building blocks necessary for language, including the capacity for categorical perception, combinatorial structure, semanticity, displacement, advanced working memory and general learning abilities, our closest relatives in the animal kingdom have not managed (or needed) to go beyond a small repertoire of vocalisations. Humans, on the other hand, seem to be alone in the ability to coordinate all the necessary ingredients for language. With our inherited ability to isolate and combine sounds, we went further and combined them creatively, as the need arose, to form an abundance of words that allowed us to refer to an ever-increasing number of concepts. Once lexical features were in use, words could be combined to produce an endless variety of messages. Through a further process of "abstraction", we developed complex grammar, involving embedding of structure and multiple other devices which allow us to express an infinite number of notions and intentions. Up to now, there is very little evidence of such creativity elsewhere in the animal world.

Cerebral connectivity: Neural connections between different areas of the brain. Modulated by learning and the environment.

According to Hauser, Chomsky and Fitch [1], the major systems of language were brought together suddenly by an undefined "rewiring" of the brain. This is in direct contrast to the adaptationist account, which holds that increased connectivity between multiple components was not saltational but a gradual process, with constant refining of interfaces between all modules, allowing for the co-evolution of phonology, syntax and semantics. We shall investigate this scenario further in Chapter 10.

Hockett added quite a few more defining features of language to his list until there were sixteen. He claimed primate communication utilises nine of these, while the rest are reserved just for humans. As we have seen, many of the features Hockett claimed were unique to humans have roots in the animal world. Nevertheless, he was correct in noticing that only human language exhibits full duality, which of course is a prerequisite for productivity. Up to now, there is very little evidence of such creativity elsewhere in the animal world. Although Hockett's behaviourist approach has since been challenged, it did in fact lay the foundations for evolutionary linguistics by looking at language from a biologically orientated, comparative perspective. By comparing the communication systems of different animals and humans, he was not only able to find features of communication that were unique to human language, but also highlight the commonalities among all human languages, which we shall now explore.

9.5 Greenberg's Language Universals

> **Language universals**: Set of structural features shared by all human languages, pioneered by Joseph Greenberg in the 1960s and 70s. Mostly concerned with the categories of syntax (words, morphemes) and how these are organised into phrases and sentences.

The quest to identify **language universals**—structural features shared by all human languages—was taken up in earnest in the 1960s and 70s by the American linguist Joseph Greenberg. From a study of some 30 languages, he derived a set of 45 basic language universals, such as:

> In declarative sentences with subject and object, the dominant order is one in which the subject precedes the object.

In English, the dominant order is SVO (*I saw Harry*) but OSV is also possible (*Bill I didn't see*). Today, we know there are languages with the dominant order VOS (e.g. Fijian), OSV (e.g. Xavante) and OVS (e.g. Hixkaryana). Linguists Nicolas Evans and Stephen Levinson [22] point out that among the 7000 or so languages spoken around the world today, there will be exceptions to nearly every universal. They claim there are merely strong tendencies at best, rather than universals. This view has led some to suggest a narrower list of "absolute" universals that apply to every known (spoken) language, such as:

> All languages have units of sound including consonants and vowels.
> All languages combine sounds to create meaning.
> All languages make use of stored chunks of sounds as words.
> All languages have words for objects and actions.
> All languages combine words to create meaning.
> All languages can say who did what to whom.
> All languages can ask questions.
> All languages can negate sentences.

This narrower list of universals stresses the "unity" rather than the "diversity" of languages, focusing on the biological constraints placed on all languages rather than the culturally determined idiosyncrasies of specific languages. This idea is mirrored in Chomsky's distinction between **I-language** (language as a mental object in a universal form) and **E-language** (language as externalised and transmitted by the community). As a nativist and "internalist", Chomsky believes that I-language is encoded in our genes and that E-language is simply generated from I-language via the sensorimotor system for the purpose of interaction. The opposite "externalist" view (held by the culturalists) is that language starts with E-language from which I-language is derived!

> **I-language**: Intrinsic linguistic knowledge held in the mind of a speaker. Can be observed indirectly through a speaker's "intuitions" about what constitutes a possible sentence in their language.

> **E-language**: Language as externalised and transmitted by the community. The tangible, observable expression of language that can be recorded and studied by scientists and linguists.

Chomsky's focus on the mental source of language led to the "cognitive" turn in linguistics, and with rapid developments and new techniques in evolutionary biology, there has been a systematic move to identifying precursors to human language in the cognitive abilities of our primate relatives. **Animal cognition** continues to be an active and evolving area of research in language evolution studies. We will now delve into two notable areas of interest, the conceptual system and social cognition.

> **Animal cognition**: Mental capacities of non-human animals. Central source of information relevant to the evolution of human cognition and, thus, language.

9.6 The Conceptual System

Most cognitive scientists today would broadly agree that communication is split into two distinct components: the first, **conceptualisation**, involves making a mental representation of an entity, and the second, lexicalisation, involves linking the mental representation to the appropriate signal in the language spoken. In the last 50 years, it has been recognised that humans and animals share much of the first (non-linguistic) component of communication. According to Fitch [19], animal cognition research is so advanced that, for many phenomena, there is little reason to doubt that mental representations exist in animals and can be recalled and manipulated.

> **Conceptualisation**: In cognitive science, the process of making mental representations of objects and concepts and mapping relationships between them.

The literature on primate behaviour in the wild gives us good grounds to assume that at least some of the basics of the human conceptual system are present in other primates. Comparative studies have shown, for example, that other species display object **categorisation**, i.e. they are able to group sensory stimuli into meaningful categories based on shared characteristics. For instance, it is common for gorillas to select different types of food based on colour and texture. Moreover, researchers from the University of Rennes have found that the alarm calls of male guenon monkeys in Western Africa vary depending on the predator. Upon hearing the noise for "eagle", the other monkeys take refuge in lower branches to avoid threats coming from the sky, whereas the sound for "leopard" sends them to the treetops [23]. Another group of researchers at the Department of Comparative Language Science of the University of Zurich have found that chimpanzees have different calls for different types of food. Upon hearing these calls, the chimpanzees search for food in the appropriate place [24].

Categorisation: Ability to group objects or elements that are alike (including syntactic constituents).

Japanese researchers Haruka Fujita and Koji Fujita [25] go as far to hypothesise that the human mental lexicon evolved from the primitive conceptual elements found in monkey calls and that these elements became externalised in humans' earliest words. However, there are important differences between words and monkey calls. Firstly, the number of signals used in monkey calls is relatively small. Words, on the other hand, have a larger number of building blocks (speech sounds) which can be combined creatively to produce a large vocabulary. Secondly, monkey calls are restricted to objects and events experienced in the present. Due to substantial extensions to the human conceptual system and the capacity for displacement, words can represent situations that are completely unrelated to the current situation. Thirdly, monkey calls have a limited function related to warning and food searching, whereas words can refer to any situation imaginable.

The one module often claimed to be exclusive to the human species is syntax. However, field researchers have found that some primates not only make use of basic categorisation but also **linear sequencing** in their calls, i.e. the sounds they make can be combined in a linear fashion (one after the other) to form new calls. For example, when the alarm call of a guenon monkey is followed by "oo", scientists believe it is a way to reduce the urgency of the call; the other monkeys remain alert but do not hide [26]. The researchers at Zurich have found that chimpanzees make one type of alarm call when they encounter danger. When they request assistance, they make another call. If they encounter a danger and need assistance, they combine the two calls. The fact that there are simple forms of syntax in our closest relatives is an indication that the syntax found in human languages might have its roots in the early forms demonstrated in chimpanzees [24].

> **Linear sequencing**: Organising principle whereby elements (including syntactic constituents) are ordered sequentially (one after the other) rather than hierarchically.

In his *Précis of Foundations of Language* (2003), gradualist Ray Jackendoff suggests linear sequencing is a necessary first step towards hierarchical structure in language. He argues that linear word order involves straight phonology-to-semantic mappings, which is ultimately of limited use. The ability to organise words and phrases into nested, recursive patterns would have allowed for more flexible and nuanced communication:

> "One can imagine the capacity for modern syntactic structure evolving *last*, as a way of making more complex semantic relations among the words of utterances more precisely mappable to linear word order in phonology. That is, syntax develops in evolution as a refinement, a 'super-charger' of a preexisting interface between phonology and semantics." (Jackendoff, 2003, 664–665) [27]

Scientists are currently researching the fundamentals of grammar and word order among chimpanzees, gorillas and orangutans at Basel Zoo. Their hypothesis is that the origin of grammar lies in **causality** and the ability to perceive events in the world in a structured way. It seems causality is already anchored in our pre-language mindset: we perceive the world in the categories of agent and patient. We have, so to speak, causal detectors in our heads, and this pattern of perception is reflected in language. The question is whether there are also such "detectors" in the heads of our closest relatives in the animal kingdom. Biologist and cognitive scientist Professor Klaus Zuberbühler from the University of Neuchatel is leading the research in Basel. Using an eye-tracker, the aim is to see if apes observe scenes in the same way as humans do, i.e. by first observing the agent, then the patient, and then switching back and forth—all within a few milliseconds. If the results confirm this, Zuberbühler believes it should be concluded that the ape brain is in principle ready for grammar [28].

> **Causality**: Cause-effect relations. Causal cognition in humans is central to explaining sentence structure. Sentences express events, which are based on who does what to whom.

Fujita and Fujita [25], in their conviction that "every sub-function of human language keeps evolutionary continuity with other species' cognitive capacities", believe that hierarchical syntax did not appear suddenly in humans but evolved from the motor action planning seen in chimpanzee tool use. Our ability to build up words into sentences, they say, has its roots in the more general ability of chimpanzees to use multiple actions on multiple objects to perform tasks. The nut-cracking behaviour of chimpanzees can be represented as follows, based on the model of Matsuzawa [29], whereby a stone wedge is used to stabilise a stone block on which a nut is placed, which is to be crushed with a hammer stone (Fig. 9.2).

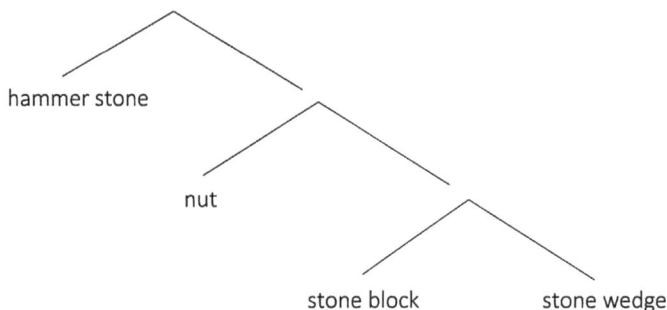

Fig. 9.2 Chain of thought involved in chimpanzee nut-cracking (based on Matsuzawa 1996: 204) [29]

Nevertheless, grammar is not limited to ordering concepts (sequentially or hierarchically) but uses a multitude of other devices related to thought and language itself, such as pronouns (to cut redundancy), articles (to signify which information is old and new), quantifiers (to specify which objects and how many), tense (to note the time of an action) and aspect (to note the duration of an action). It also offers the means to form questions, negatives, commands, demands, promises, threats, etc. Language in its entirety does an amazing job not only of representing who does what to whom, but also conveying notions of space, time, motion, matter, social relations, negation, conditions, hypothetical situations and goals. It is undoubtedly qualitatively different from anything found in the animal world.

9.7   Social Cognition

We will now turn to some evidence for the beginnings of human **social cognition** in non-human primates. Although apes are somewhat lacking in vocal skills, the last two decades have revealed that great apes are rather adept at understanding the goals and intentions of others, which they use to their advantage for survival. For example, Yamamoto et al. (2012) [30] found that when chimpanzees see an individual trying to reach for an out-of-reach object, they often recognise the goal and help the individual to achieve it. Understanding the mental states of others is relevant to the skills of social learning, on which language is based.

There is now evidence for the use of social cognition specifically in the communication systems of our primate relatives. Researchers from the University of Rennes have found that female guenon monkeys, who are responsible for keeping their families together, make contact calls, and these calls vary slightly according to social influences. For example, two females who are friends will copy each other. The scientists have also discovered conversation rules: the monkeys learn not to interrupt each other and to let the elder females speak first. It has been demonstrated that the young monkeys interrupt others twelve times more often than the adult monkeys do

[31]. These rules of conversation seem to point to a certain degree of mind-reading or "theory of mind" (see Sect. 3.1), since they suggest a level of social awareness and consideration for the perspectives of others. However, according to British cognitive scientist Thom Scott-Phillips (2014) [32], there is one big difference between the social communication of animals and humans: only humans have the ability to communicate intentions "ostensively". The signals animals use simply reflect pre-established social norms and therefore do not explicitly reflect their own mental state to the recipient, which would depend on a more sophisticated theory of mind and require a more complex code of symbols [32]. Nevertheless, the findings contribute to our understanding of the evolutionary origins of certain aspects of social behaviour and the development of communication skills in humans. Moreover, they emphasise the fact that language serves multiple functions, including the reinforcement of social bonds and establishment of group cohesion.

Social cognition: Mental capacities used in social interaction.

Figure 9.3 shows the evolutionary development of the brain and resulting traits relevant to the development of language. It distinguishes between shared traits (those handed down to humans from other species) and derived traits (those that developed only in humans), although many derived traits are effectively modifications of shared traits. Birds occupy a special twig on the tree of life (and therefore do not appear in the chart below). Emerging 165–150 million years ago as an offshoot of the theropod clade of dinosaurs (which developed some 230 million years ago from reptiles), birds have an independent evolutionary history involving the development of several traits which reappeared much later in humans for the purpose of language (including advanced vocal control, combinatorial structure, vocal imitation, vocal learning and cultural transmission).

9.8 Chapter Summary

A theory of language origins should go far beyond looking at the external structure of the communicative system. It should also take an internal perspective and focus on the biological and cognitive adaptations that allowed humans to use and acquire language. Hockett's classification system, while useful for describing and comparing the external communication systems of humans and animals, remains "of limited use as a theoretical framework for evolutionary linguistics" since it focuses on "the communicative code itself rather than the cognitive abilities of its users" [11]. Nevertheless, Hockett did in a sense establish a basis for evolutionary linguistics by looking at language from a biologically orientated, comparative perspective. The cognitive turn in linguistics led to the recognition that the principles of language have their roots in multiple cognitive domains found in humans and other animals,

Time (mya)	Emergent organism	Shared features (handed down to humans)
800	Single-cell (e.g. amoeba)	Ability to sense and respond to environment
580	Multicellular (e.g. sponge)	Ability to sense and respond to other cells
540	Specialised-cell (e.g. jellyfish)	First neurons; non-centralised nerve net
525	Chordata (e.g. flatworm)	Centralised nervous system
350	Amphibians (e.g. frog)	Basic brain; centralised body functions/breathing
312	Reptiles	Development of brainstem and cerebellum Ability to cooperate, forage and defend territory
250	Mammals	Appearance of hindbrain Involuntary vocalisation Development of cerebrum Emotions and bonding
85	Primates	Development of cortex Enhancement of sensory perception Categorical perception
40	Monkeys	Development of neocortex Use of calls to refer to concepts Social cognition Use of influence, rank and deception
23	Apes	Development of prefrontal cortex Coordination of sensory input and motor control Combining of calls (proto-syntax) Causal cognition Future planning/displacement Problem-solving Cultural transmission
		Derived features (unique to humans)
6	Humans	Enlargement of brain Greater cerebral connectivity Advanced voluntary control Extended conceptual system Naming insight Combinatorial phonology/productivity Complex motor memory Combining of concepts into complex thoughts Extensive vocal imitation and learning Advanced working memory and long-term memory

Fig. 9.3 Tracing the evolution of brain and language: From single cells to humans

and since the 1990s, the comparative study of cognition across species has been complemented by the empirical factor, turning the question of language origins into a scientific enterprise.

Comparative studies have not only given us insights into the evolutionary path of human language but have also highlighted the fact that language is unique compared to other animal communication systems. In particular, the fact that language is externalised and transmitted culturally has turned it into a highly developed cultural tool that can adapt to the needs of its users. Today, it is becoming increasingly clear that, in addition to the cognitive underpinnings of language, the social environment and cultural evolution of language are factors that cannot be neglected in evolutionary models of language.

In the next and final chapter, I will bring together all the evidence from the behaviourists, nativists, adaptationists and culturalists to explain how a sophisticated communication system involving a unique combination of features found nowhere else in nature came to be.

References

1. Hauser, M. D., Chomsky, N., & Fitch, W. T. (2002). The Faculty of Language: What Is It, Who Has It, and How Did It Evolve? *Science, 298,* 1569–1579. https://doi.org/10.1126/science.298.5598.1569
2. Pinker, S., & Jackendoff, R. (2004). The faculty of language: What's special about it? *Cognition, 95*(2), 201–236. https://doi.org/10.1016/j.cognition.2004.08.004
3. Hockett, C. (1960) The Origin of Speech. *Scientific American, 203,* 88–111. https://doi.org/10.1038/scientificamerican0960-88
4. Stein, J. F. (2003). Why did language develop? *International Journal of Pediatric Otorhino-laryngology, 67*(1), 131–135. https://doi.org/10.1016/j.ijporl.2003.08.011
5. Zuidema, W., & de Boer, B. (2018). The Evolution of Combinatorial Structure in Language. *Current Opinion in Behavioural Sciences, 21,* 138–144. https://doi.org/10.1016/j.cobeha.2018.04.011
6. Kuhl, P. K., & Miller, J. D. (1975). Speech perception by the chinchilla: Voiced-voiceless distinction in alveolar plosive consonants. *Science, 190*(4209), 69–72. https://doi.org/10.1126/science.1166301
7. Fitch, W. T. (2018). The Biology and Evolution of Speech: A Comparative Analysis. *Annual Review of Linguistics, 4*(1), 255–279. https://doi.org/10.1146/annurev-linguistics-011817-045748
8. Beecher, M. D. (2021). Why are no animal communication systems simple languages? *Frontiers in Psychology, 12*(602635). https://doi.org/10.3389/fpsyg.2021.602635
9. Leroux, M., & Townsend, S. W. (2020). Call combinations in great apes and the evolution of syntax. *Animal Behavior and Cognition, 7*(2), 131–139. https://doi.org/10.26451/abc.07.02.07.2020
10. Seyfarth et al. 1980, as cited in Aitchison, J. (2000). *The Seeds of Speech.* Cambridge University Press.
11. Wacewicz, S., & Zywiczynski, P. (2015). Language evolution: Why Hockett's design features are a non-starter. *Biosemiotics, 8,* 29–46. https://doi.org/10.1007/s12304-014-9203-2
12. Cuskley, C. (2020). *Language Evolution: A brief overview.* PsyArXiv. https://doi.org/10.31234/osf.io/3y98j

13. Deacon, T. (2010). A role for relaxed selection in the evolution of the language capacity. *PNAS, 107*(2), 9000–9006. https://doi.org/10.1073/pnas.0914624107

14. Inoue, S., & Matsuzawa, T. (2007). Working memory of numerals in chimpanzees. *Current Biology, 17*(23), 1004–1005. https://doi.org/10.1016/j.cub.2007.10.027

15. Marler, P., & Sherman, V. (1985). Innate differences in singing behaviour of sparrows reared in isolation. *Animal Behaviour, 33*(1), 57–71. https://doi.org/10.1016/S0003-3472(85)80120-2

16. Gruber, T., Muller, M. N., Strimling, P., Wrangham, R., & Zuberbühler, K. (2009). Wild Chimpanzees Rely on Cultural Knowledge to Solve an Experimental Honey Acquisition Task. *Current Biology, 19*(21), 1806–1810. https://doi.org/10.1016/j.cub.2009.08.060

17. Pepperberg, I. M. (1999). *The Alex Studies: Cognitive and communicative abilities of grey parrots.* Harvard University Press. https://doi.org/10.2307/j.ctvk12qc1

18. Osvath, M., & Osvath, H. (2008). Chimpanzee (Pan troglodytes) and orangutan (Pongo abelii) forethought: Self-control and pre-experience in the face of future tool use. *Animal Cognition, 11*, 661–674. https://doi.org/10.1007/s10071-008-0157-0

19. Fitch, W. T. (2019). Animal cognition and the evolution of human language: Why we cannot focus solely on communication. *Philosophical Transactions of the Royal Society, Series B, Biological sciences, 375*(1789), 20190046. doi:https://doi.org/10.1098/rstb.2019.0046

20. Lieberman (1984), as cited in Pinker, S., & Jackendoff, R. (2004). The faculty of language: What's special about it? *Cognition, 95*(2), 201–236. https://doi.org/10.1016/j.cognition.2004.08.004

21. Lieberman (1984, 1992), as cited in Aitchison, J. (2000). *The Seeds of Speech.* Cambridge University Press.

22. Evans, N., & Levinson, S. (2009). The Myth of Language Universals: Language Diversity and Its Importance for Cognitive Science. *Behavioral and Brain Sciences, 32*(5), 429–448. https://doi.org/10.1017/S0140525X0999094X

23. Ouattara, K., Lemasson, A., & Zuberbühler, K. (2009a). Campbell's monkeys use affixation to alter call meaning. *PLoS ONE, 4*(11), e7808. https://doi.org/10.1371/journal.pone.0007808

24. Gull, T. (2021). Mensch und Schimpans. *UZH Magazin, 2*(21), 32–37.

25. Fujita, H., & Fujita, K. (2021). Human language evolution: A view from theoretical linguistics on how syntax and the lexicon first came into being. *Primates, 63*(5), 403–415. https://doi.org/10.1007/s10329-021-00891-0

26. Ouattara, K., Lemasson, A., & Zuberbühler, K. (2009b). Campbell's monkeys concatenate vocalizations into context-specific call sequences. *PNAS, 106*(51), 22026–22031. https://doi.org/10.1073/pnas.0908118106

27. Jackendoff, R. (2003). Précis of Foundations of language: Brain, meaning, grammar, evolution. *Behavioral and Brain Sciences, 26*, 651–707. https://doi.org/10.1017/S0140525X03000153

28. Nickl, R. (2021). Aktionfilme für Affen. *UZH Magazin, 2*(21), 38.

29. Matsuzawa, T. (1996). Chimpanzee intelligence in nature and in captivity: Isomorphism of symbol use and tool use. In J. Goodall, J. Itani, & W. Foundation (Authors) & W. McGrew, L. Marchant, & T. Nishida (Eds.), *Great Ape Societies* (pp. 196–210). Cambridge: CUP.

30. Yamamoto, S., Humle, T., & Tanaka, M. (2012). Chimpanzees' flexible targeted helping based on an understanding of conspecifics' goals. *Proceedings of the National Academy of Sciences, 109*(9), 3588–3592. https://doi.org/10.1073/pnas.1108517109

31. Lemasson, A., Gautier, J.-P., & Hausberger, M. (2004). Vocal similarities and social bonds Campbell's monkey (Cercopithecus campbelli). *Comptes Rendus Biologies, 326*(12), 1185–1193. https://doi.org/10.1016/j.crvi.2003.10.005

32. Scott-Phillips, T. C. (2014). *Speaking Our Minds.* Palgrave Macmillan.

Chapter 10
Conclusion: Putting It All Together

It is surely clear by now that there is no single factor responsible for triggering language in humans. Some theories regard language primarily as a mental object, an internal system of thought, while others stress the importance of language as a means of exchanging information for group survival. Yet other theories propose that language arose for social reasons, to form alliances and social ties, or else for antisocial reasons, to deceive and outwit rivals. It is likely that a combination of factors, including the functional, social and cognitive, contributed to the emergence of language in humans. Language is both a cognitive and an interactional phenomenon, and its design reflects this fact, with multiple systems superimposed on each other. This chapter reviews the various aspects involved in the evolution of language and considers how we can better synthesise the ever-increasing quantities of data from multiple disciplines.

In our exploration of the origins of language, it was first necessary to describe our object of inquiry, language itself, and this required a thorough dissection of its structure and organisation. In Chapter 2, in the structuralist tradition, we broke down language into its basic components (sounds, words and morphemes) and saw how these combine so as to produce a language that is creative and boundless. In Chapter 3, we turned to the purpose of language and saw it had various functions, including the expressive, directional, social and informational. All of these functions feature in early theories of language origins. However, being merely conjectures, some more fanciful than others, most of these hypotheses were discarded by "more serious researchers" in the latter half of the nineteenth century. Nevertheless, they did provide some interesting insights, particularly the idea that language is multimodal and could have started out being more gestural than verbal. Another major insight was the involvement of the musical system, suggesting that the evolution of language is highly multifactorial.

Then we moved on to environmental factors (both natural and social) affecting the development of language. In Chapter 4, we saw how adaptation to the physical environment gave rise to physical and cognitive traits in humans that were essential

© The Author(s), under exclusive license to Springer Nature Switzerland AG 2024 137
J. Dornbierer-Stuart, *The Origins of Language*,
https://doi.org/10.1007/978-3-031-54938-0_10

for language. Our timeline began with the major tectonic shifts that dried up the forests of East Africa and dramatically changed the environment. Early humans took to hunting and eating meat and standing on two feet to better scan the land for game and enemies. As well as freeing the hands for tools and weapons, the upright stance altered the shape of the mouth and throat, allowing humans considerable articulatory agility when vocalising. It also narrowed the hips, making childbirth problematic, with the result that children were born earlier, with heads and brains developing substantially after birth. This caused children to be more dependent on elders, which encouraged stronger social ties and social skills. In addition, being born "prematurely" meant brains were in a relatively undeveloped state, remaining highly plastic and sensitive to environmental stimuli for longer, which was crucial for language learning.

In Chapter 5, we turned to the social and cultural conditions likely to have influenced the development of language. Because humans are social animals, there would have been a need to cooperate for the procurement of food and raising of offspring, and this would have required efficient communication. In addition, larger groups of humans would have benefitted from forming friendships and hierarchies to keep order and peace, and this could have been achieved vocally, through gossip and storytelling. Then we explored the notion that language was an external cultural object and self-evolving. Through social interaction and cultural transmission, language would have developed further, and with language becoming more and more productive, this would have created a linguistic environment to which humans had to adapt. Lastly, in Chapter 5, we were introduced to Kirby's iterated learning model, which emphasises the key role cultural transmission plays in shaping linguistic structure.

According to evolutionary biologist W. T. Fitch [1], the biological evolution of the capacity for language and the cultural evolution of language are still considered by some, e.g. Christiansen and Chater [2], to be "mutually exclusive competing explanations". Biological accounts emphasise that language evolved as an adaptation, providing selective advantages for communication and cooperation among early humans. The idea here is that, over time, natural selection favoured individuals with genetic predispositions for language-related abilities, leading to the development of language as a complex cognitive function. The challenge for purely biological accounts is to account for the incredible diversity of languages and the speed at which language evolved, known as "Darwin's problem". Cultural evolution theories, on the other hand, focus on the role of social and cultural dynamics in shaping linguistic behaviour and can thus provide a mechanism for explaining how and why languages change so rapidly, taking pressure off biological explanations. The challenge for cultural accounts is to link cultural processes to their biological hosts.

At this point, I would like to note the role Saussure played in connecting the cultural system of language to cognition. The father of structuralism, Saussure is primarily known for having portrayed language as a closed system of arbitrary signs which derive their significance from their relations to other signs. However, Saussure also suggested that since there was no inherent connection between the sound pattern of a word and its meaning, the association between the two was established psychologically, i.e. in the mind of the language user. Chomsky expanded on this and proposed that the structural principles of grammar are hardwired in the human brain, and that this provides an initial state and foundation for children's language learning. His theory of Universal Grammar not only revolutionised linguistics, turning it into a "science of the mind", but also contributed to various disciplines concerned with the study of the mind and intelligent behaviour. In particular, his notion of modularity of the mind, suggesting that the human mind consists of distinct, specialised modules or systems, has informed research in cognitive psychology and neuroscience.

While Chomsky may be considered one of the founders of cognitive science, it is his student, Ray Jackendoff, who perhaps more than any other linguist has worked to integrate linguistics into the cognitive sciences. In the 1970s and 80s, Jackendoff made breath-taking analyses of how we use grammar to represent truly abstract concepts involving space, time, motion, location, causation, goals and social relations, creating as it were a blueprint for the structure of thought, which he called "mental anatomy". This was an important step in linking syntax with semantics and realising that the latter was the cradle of the former instead of the other way round. In the 1990s, Jackendoff turned his attention to the "function" of such abstract structure. He made the common-sense assumption that language arose gradually in the interest of enhancing communication, and this is what prompted his interest in integrating linguistic theory into biological evolution.

Following the adaptationist doctrine, Jackendoff views language as a complex adaptation evolving in stages. Based on Darwinian natural selection, the theory implies that verbal abilities play a significant role in both the reproductive success of individuals and the success of our species as a whole. Clearly, evolutionary explanations for the origins of language need to postulate genetic changes required for language. Unfortunately, we still know relatively little about the relevant mutations in the evolution of language. The few that have been postulated concern the neural control of vocalisation (not grammar!). According to Fitch [3], by far the best-known gene involved in human vocal control is FOXP2, which is found in all vertebrates but modified in humans. It is believed to play a pivotal role in improving oral motor sequencing skills in humans. For other components of language, including the perceptual (e.g. audition), the conceptual (e.g. categorisation and ordering of concepts), vocal learning and social cognition, it is extremely difficult to demonstrate encoding by genes. According to geneticist Simon Fisher [4], the study of language genes is challenged by the fact that no gene is limited to a single function, but instead participates in more than one process. Also, genes do not operate in isolation but interact with other genes in networks and complexes.

However, Fisher believes that as our understanding of gene networks continues to grow, more light will be shed on the role of specific genes in further areas of language processing. Already, virtually complete genome sequences from modern and ancient humans have revealed that the two coding changes in FOXP2 that distinguish the human sequence from that of chimpanzees were already present in our Neanderthal cousins. However, further in-depth comparisons of Neanderthal and modern human versions of the gene point to human-specific changes that might affect the way that the gene is regulated. This work highlights the fact that evolutionary linguistics has progressed beyond rationalist theorising and is now open to empirical testing.

In Chapter 7, we saw how psycholinguistics, complemented by the findings of neurolinguistics, has done much to unravel the mysteries of language processing in humans. While linguists are good at describing the various systems of language available to users—the sound system (phonology), the combinatorial system (syntax) and the conceptual system (semantics), psycholinguists are good at explaining the processes involved in using language (i.e. speech comprehension and speech production). In addition, neuroimaging can indicate which brain structure and neural pathways support which aspects of language. Neuroscientists are increasingly able to explain the neural networks behind the main systems of language and how they interact during language use. By comparing the brains of humans and other primate species, it is possible to make assumptions about which structures and pathways are at the root of the human capacity for language.

In Chapter 8, we saw that children are born with a subtle mix of cognitive mechanisms, including categorical perception, imitation and general learning skills, that provide a biological framework for language to develop in the individual. We saw some evidence for language learning being innately guided, and learnt about Chomsky's set of structural principles (Universal Grammar) that hold for all languages, which he later reduced to a single computational operation (Merge). However, in a theory of language evolution, it is not enough to consider language as a psychological object that belongs to the individual since it is ultimately an externalised behaviour that needs to be learnt from others. Without the ability to learn from others, language would not exist.

In Chapter 9, we turned to the comparative approach to find out which aspects of language are likely to be inherited from other species and which evolved later and only in humans. The most unique features of human language are arguably vocal agility and cerebral connectivity, which would have resulted from (in evolutionary terms) relatively recent adaptations to the vocal tract and neural circuitry. By contrast, comparative studies of the cognitive abilities across species have revealed that many sub-components of the major systems in language (phonology, semantics and syntax) are shared with other species, e.g. categorical perception, conceptualisation, categorisation and sequencing) and therefore have more ancient evolutionary roots.

Just as we can observe sub-components of the major systems of language in animals, we can also find evidence of connections between these systems in the animal world. For example, in birdsong there is patterning of sound, i.e. the combinatorial (syntactic) system (let's call it A) is linked to the sound (phonological) system (B). However, the songs have no referential meaning, i.e., the sound system (B) is not linked to the conceptual (semantic) system (C). Thus, birdsong features bare phonology (AB) but not semanticity (AC). By contrast, in ape calls, different sounds refer to different objects, i.e. the sound system (B) is linked to the conceptual system (C), but there is no significant patterning of sounds, i.e., the sound system (B) is not linked to the syntactic system (A). Thus, primate calls feature semanticity (AC) but not bare phonology (AB). The very interesting "Integration Hypothesis" (Miyagawa 2017) [5] holds that, at some point in recent evolutionary time, the bare phonology found in birdsong (AB) and the semanticity found in primate calls (AC) integrated

A   **B**   **C**

PATTERN SYSTEM   SOUND SYSTEM   CONCEPTUAL SYSTEM

A B   **B C**

BARE PHONOLOGY   SEMANTICITY
(BIRDSONG)   (APE CALLS)

A B C

PROTO-WORDS
(HUMAN LANGUAGE)

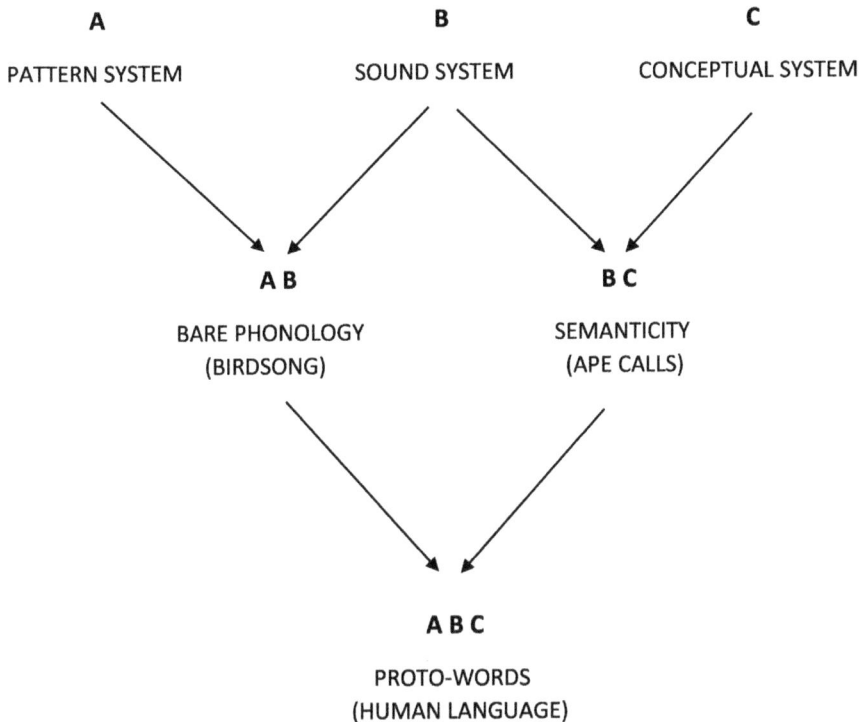

Fig. 10.1 Integration of combinatorial structure and semanticity in human language

relatively abruptly and uniquely in humans (in line with saltationism) to produce
a three-way system (combinatorial phonology) where sounds were combined into
elements which have meaning (ABC), producing proto-words (see Fig. 10.1).

According to this view, the syntactic system, if it is found in birds, must have
appeared early on in evolutionary history, prior to humans. Likewise, the conceptual system is not unique to humans. This suggests that the novel step in human
language was not syntax nor semantics but the ability to link bare phonology (AB)
to semanticity (AC) in proto-words, i.e. representations encoding phonological,
syntactic and semantic properties simultaneously. Only once lexical features were
in use could words have been combined into phrases and sentences to produce an
endless variety of messages. Words had to come first. Chomsky's premise that Merge
is the only mechanism that would have needed to evolve for language to evolve must
be considered false. It would have been in a later step that words were combined
into sentences, possibly starting with simple compounding to denote objects (e.g.
cry-baby), followed by linear sequencing to denote events (e.g. *baby sleep*), and
eventually hierarchical structure.

Chomsky's theory was again adapted and, in 2002, Hauser, Chomsky and Fitch
came up with their influential model of language. It conceived of the human language

faculty as a system consisting of three independent components: a "conceptual-intentional system", which relates to meaning and interpretation, a "sensorimotor system", relating to the perception and production of linguistic signals, and the "computational system", responsible for recursion. While persisting in the claim that the third component is unique to humans, the researchers did suggest that humans and animals share a diversity of computational resources for domains other than communication (e.g. number, navigation), but that there has been a "substantial evolutionary remodelling since we diverged from a common ancestor some six million years ago" that allows for recursion for communication [6].

In later modifications (Berwick and Chomsky, 2011 [7], 2016 [8]), it was claimed that the computational system was first optimised for interaction with the conceptual-intentional system to form an "internal language of thought" (ILOT) that is unique to humans. Subsequent connection of the computational system with the sensorimotor system enabled the ILOT to be externalised for interaction and communication, causing diversity among human languages. This fits in much better with the idea that our closest relatives, chimps, have many of the cognitive abilities for language but have limited vocal abilities. It has been suggested by Jackendoff [9] that connection of the computational system with the conceptual-intentional system to form the ILOT is a "pre-stage" to grammar, with rudimentary principles, including ordering. Although in no way constituting fully developed grammar, these principles are a necessary step to full language.

Fujita and Fujita [10] came up with an interesting counter hypothesis (the "Disintegration Hypothesis") that suggests that while some non-human primate calls have been shown to exhibit both semanticity and combinatorial structure, these two systems remain fused or "undifferentiated", i.e. animal communication cannot separate referential content from emotional content. In human language, the two systems became differentiated into lexical categories (expressing referential information) and functional categories (expressing information relating to actual circumstances). Fujita and Fujita suggest that the disintegration took place gradually, during the transition from protolanguage to human language.

We saw above that without words, syntactic structure could not have arisen. If we think logically, words (which encode semantic, phonological and syntactic properties simultaneously) could not have come into being without syntactic structure. Ray Jackendoff [9] addresses this problem in his "Parallel Architecture" of language, which can be used as a vehicle to explain the evolution of the language capacity. It holds that the three main systems of human language—phonology, syntax and semantics—are equally well-formed with their own distinct internal computational rules, and all are interlinked or correlated by interface components (see Fig. 10.2).

For example, as we saw in Sect. 2.1, phonology has a hierarchical structure of its own that is relatively independent of syntactic structure, although not unconnected. At the bottom level of phonological structure, we have speech sounds, which are grouped into syllables. Syllabic structure itself is hierarchical, with embedding. At the centre of a syllable is a nucleus (usually a vowel) and on the edge are consonants. Syllables group into phonological words, which group into phonological phrases. These hierarchical units are not built out of syntactic primitives but are intrinsically

Fig. 10.2 Jackendoff's "parallel" view of language

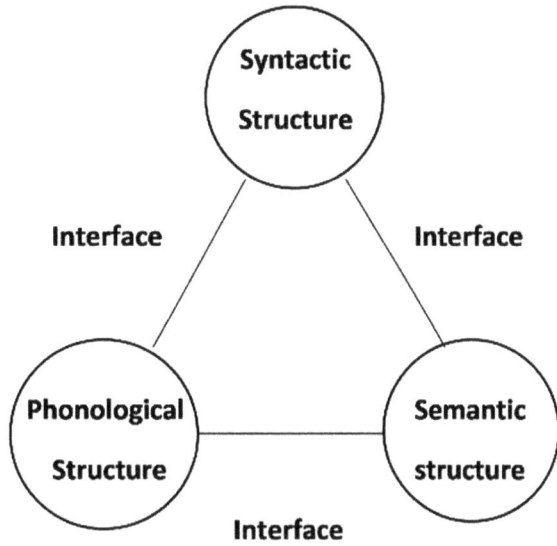

phonological [11]. For example, *Lisa's (*as in *Lisa's a doctor)* is clearly one phonological word that corresponds to two separate syntactic constituents [11]. Similarly, phonological phrases do not coincide exactly with syntactic phrases, as the following example shows:

Syntactic bracketing:
[This] [is [the cat [that chased [the rat] [that ate [the cheese]]]]]]

Prosodic bracketing:
[This is the cat][that chased the rat][that ate the cheese] [11]

Meaning, too, has its own hierarchical structure, with meaningful elements (morphemes) embedded in words, which are embedded in grammatical and semantic categories (e.g. agent, patient) in sentences, which themselves are assigned to functional categories (e.g. request, apology). These semantic layers not only need to interact with each other through interfaces, but also with the layers of phonology and syntax. In this view, words can be regarded as a special type of interface that combines pieces of phonological, syntactic and semantic structure. Thus, in the words of Jackendoff, "language does not consist of a lexicon *plus* rules of grammar; rather, lexical items are among the rules of grammar" [11].

According to Jackendoff, the conception of interfaces within language aligns perfectly with theories of speech processing. In Chomsky's "syntactocentric" view of language, the phonological and semantic structures cannot be determined without first establishing the syntactic structures (see Fig. 10.3).

Fig. 10.3 Chomsky's
"syntactocentric" view of
language

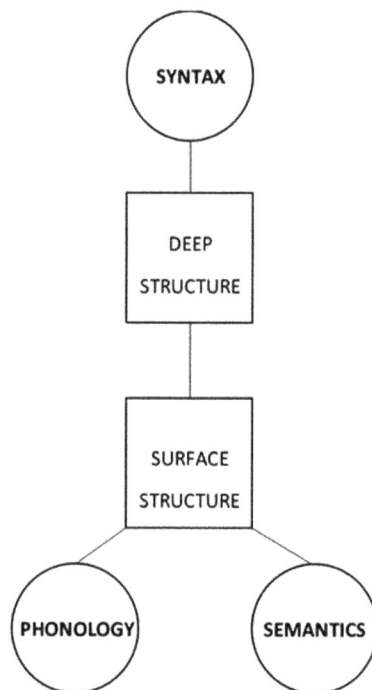

However, as we saw in Chapter 7, speech perception proceeds in one direction:

Sound → Phonology → Syntax → Meaning

and speech production in the other:

Meaning → Syntax → Phonology → Sound

Hence, it is impossible that the single directionality in the syntactocentric model can cater to both speech perception and speech production. The Parallel Architecture, by contrast, has no central module [11]. It is rather an interplay of phonological, syntactic and semantic structures and implies a highly dynamic functioning of the mind. The syntactic module, for instance, uses syntactic cues provided by words to build a hierarchical structure for the sentence, and the semantic module uses the meanings of words to establish relationships between sentence constituents. These modules interact dynamically, allowing words to shape the structure of the sentence through their syntactic and semantic properties.

According to Jackendoff, this active role of words in determining the structure of sentences is in accordance with psycholinguistic evidence [11]. Studies on sentence comprehension and production have empirically demonstrated that words are not passive units but contribute to sentence structure through their syntactic and semantic properties. For example, the placement of a word can influence the interpretation of the sentence and the relationships between its constituents—compare *She only likes pizza* with *Only she likes pizza*. According to the Parallel Architecture theory, it is words, not syntax, that shape the structure of the sentence through their syntactic and semantic properties.

I would suggest the notion of interfaces is also in tune with theories of children's language acquisition. In Chapter 8, we saw that children's language develops in a set order (first the sound system, then words and meaning, then syntax). At the same time, it is known that each developing system has a feedback effect on the previous one and influences the next one. For example, developing phonetic and prosodic knowledge assists in the discovery of words. The development of words feeds back on the realisation of phonemic contrasts. Sentence units seem to aid phonological acquisition. Syntactic structure appears to aid the development of word meaning, and vice versa. This suggests that infants are continuously integrating different forms of linguistic knowledge [12].

As for trying to relate the syntactocentric theory of language to the biological evolution of language, there is no way to build a syntactocentric theory of language such that the earlier stages are useful. Why would syntactic structure have evolved if there were no words to fit into it? In the Parallel Architecture approach, it is assumed the conceptual (semantic) system, already present in non-human primates, was the first "generative" component of language (i.e. concepts could be manipulated in the mind). According to Jackendoff [11], an early stage in language evolution would have been the association of vocalisations with concepts (to produce semanticity). A regimentation of speech sounds (producing combinatorial phonology) would have been the next "generative" component of language, forming proto-words. Then words would have been ordered, first using linear sequencing, forming proto-syntax, and then hierarchical structure to show more complex semantic relations between words [11].

So how can Jackendoff's model be related to Darwinian natural selection? The problem is to explain a pathway for incremental evolution such that earlier stages of language are all beneficial to the organism. According to Professor Ljiljana Progovac [13], mastering simple grammar would have brought about a remarkable increase in expressive abilities and would have helped in the competition for status and sex. Those individuals who had just a bit more verbal skill would have produced more offspring and thus passed on the genetic endowment supporting this ability. Progovac suggests that until we know for sure how humans and language evolved, "the only clear way forward is for linguists to formulate testable hypotheses" which can then be "subjected to interdisciplinary scrutiny and testing" [13].

Fitch [1] notes that with whole new classes of data, e.g. paleo-DNA (DNA recovered from archaic humans) and a wealth of more traditional types of data relating to animal cognition, genetics and neuroscience, we now have "increasingly sophisticated models of language evolution that make testable predictions" [1]. In his review of empirical approaches in the discipline, he puts forward his own multistage model of language evolution. In his words, his aim is "illustrative, to show that a model can be constructed that is consistent with all available data, and that makes clear testable predictions" [1]. The following scheme is my own tentative proposal for some key transitions in language evolution since our divergence from chimpanzees, with rough timelines. It incorporates elements of Jackendoff's gradualist scenario [11] and Fitch's adaptationist multistage model [1]:

Stage 1: 4.5–2.5 mya (Pliocene/Australopithecus)
Use of sounds in holistic messages along the lines of sophisticated chimp grunts and hoots; combining of sounds for display purposes only; limited semanticity through prosodic features, e.g. high pitch for danger, low pitch for contentment (as in ape calls). According to Fitch, this first stage would have also required vocal learning capacities to generate learned vocal sequences without propositional meaning.

Stage 2: 2.5–0.5 mya (Early Pleistocene/Homo habilis, Homo erectus)
Through analysis, specific sounds are used for specific concepts (semanticity). According to Fitch, this would have been propelled by the need to share detailed propositional information with close kin. With improved vocal control, greater differentiation of sounds. Through synthesis, combining of sounds (combinatorial phonology) to refer to more concepts. Consonant-vowel combinations. Proto-words: lexical items that can refer to whole situations.

Stage 3: 0.5–0.2 mya (Middle Pleistocene/Homo heidelbergensis)
Further combining of sounds to form a larger vocabulary. Verb-noun compounds and proto-syntax consisting of linear sequencing.

Stage 4: 0.2 mya (Late Pleistocene/Homo sapiens)
Elaboration of grammar, including the combining of words in a hierarchical order and grammaticalisation of lexical items.

There are at least three serious challenges to purely biological accounts of language evolution that focus on major transitions leading to complex language. Firstly, the novel steps (e.g. combinatorial phonology, proto-syntax) appear far too rapidly to be accounted for by the processes of genetic mutation and natural selection alone. The rapid evolution of language may be better explained by considering the role of social interactions and the cultural evolution of language. Secondly, these models often ignore other important components of language such as social cognition and theory of mind. Thirdly, such models ignore variability in language. According to Evans and Levinson [14], since humans are the only species to have a communication system that varies at every level, including form and meaning, it should be a primary concern to explain how we derive such diversity from a common biological core. They believe such linguistic diversity points to "remarkable cognitive plasticity", which allows us not only to use language flexibly and create new forms but also to adapt to diverse linguistic environments. This plasticity is of course ultimately the product of natural selection and perhaps the most significant adaptive capacity for language.

Thus, it could be argued that any theory of a natural language instinct must include an instinct to use and learn language, as it is through these skills that the potential of the instinct is realised. Chomsky's Universal Grammar, the hypothesised human-specific cognitive structure that generates grammatical sentences and guides children's language acquisition, has been criticised for dealing with syntax in a vacuum, without considering the learner's mind as a processor and learner of information. As an alternative to Universal Grammar, some have put forward a general learning module that uses inductive inference, without recourse to a special "language organ". According to this view, language is an "abstract" entity in the community and apt to rapid change as it is transmitted from person to person and from generation to generation.

This brings us to the culturalist perspective, which is currently enjoying more and more influence in theories of language evolution. The culturalists argue that it is not primarily genes but the process of cultural evolution that leads to structure in language, with some claiming that the genetic constraints on language mostly existed prior to the emergence of language itself. As human brains enlarged and children were born earlier, extending childhood, there was more time to develop language and pass it on culturally through learning. When language is transmitted culturally over time, it changes. According to Kenny Smith [15], if you take the process of cultural evolution to its logical limits, it can potentially offer a single, unified mechanism that can explain both language change and the genesis of language.

In Chapter 6, we looked at what causes language to change as it is transmitted across the community and from generation to generation. We saw how historical linguists Bernd Heine and Tania Kuteva [16] consider language change to be the key mechanism in the development of grammar, specifically through the process of grammaticalisation. According to Spanish linguist José-Luis Mendívil-Giró [17], the logic of grammaticalisation seems sound since historical linguistics has clearly demonstrated that grammatical forms are often etymologically derived from lexical forms. For example, the Modern English auxiliary verb *will*, used grammatically to form the future tense, is derived from the Old English lexical verb *willan*, meaning "to want". This seems to correspond with the intuitive idea that, in the evolution of human language, lexical items or content words (nouns, verbs, adjectives) emerged prior to grammatical items or function words (auxiliaries, affixes, prepositions, etc.). In other words, "grammatical categories would result from a process of abstraction of the more concrete and 'tangible' meanings of lexical words" [17]. However, Mendívil-Giró argues, "there is no evidence that the oldest languages to which we have access were 'less grammatical' than more recent languages, or that there exist today grammatical categories that did not exist in the past" [17]. He believes grammaticalisation does not create the syntactic categories of language but simply "reanalyses" them during the process of externalisation. In this view, language change is limited to superficial variation and does not alter the "core" of language. Besides, the opposite process, "degrammaticalisation", also occurs, whereby content words emerge from function words, as in *He upped his game*. Nevertheless, it could be counter-argued that grammaticalisation, is the more common direction of linguistic change.

In Chapter 6, we also came across the idea that grammar is constantly adapting itself to usage as language is transmitted "horizontally" (across the community). For example, in English, adjectives usually *precede* nouns (as in *A silly proposal*), unless the adjective is lengthened into a clause, in which case the adjective *follows* the noun (as in *A proposal too silly for words*). This is clearly to avoid processing difficulties: the head of a noun phrase (i.e. the noun) cannot be delayed too far in the noun phrase, otherwise the result is incomprehensible (*A too silly for words proposal*) [18].

Computer analysis of large corpora of natural data reveals that grammar never reaches a static end state but is constantly evolving. According to American linguist Joan Bybee [19], one of the main factors influencing the continual re-shaping of language is frequency of use. Frequency has two main effects. The first is a processing effect in which the phonetic shape of frequent words gets whittled down with use. For example, *going to*, a frequently used phrase in English, becomes phonetically reduced to "*gonna*", requiring less muscular effort during articulation. Secondly, frequency of use ensures that the most frequently used constructions are retained [19]. This explains why irregular past tense forms of frequently used verbs, e.g. *came*, *went*, *saw* and *took*, remain stable across time, whereas irregular forms of rare verbs, e.g. *besought*, tend to become regular (*beseeched*).

According to Christiansen and Chater [2], the structure of human language—the phonological, syntactic and semantic regularities of language—is not only shaped around processing needs but also general learning mechanisms, as language is transmitted "vertically" (from parents to offspring). For example, learning contains a bias towards simplicity, and this is put forward as an important factor in shaping language: linguistic systems that are simpler allow for shorter, more concise mental representations, which are easier to learn. As a result, some aspects of language remain and others decay. Christiansen and Chater also point out that grammar could not have arisen by biological adaptation since language is a "moving target" and, hence, "cannot provide a stable environment to which languages could have adapted". They conclude that language is easy for us to learn and use, not because our brains adapted to language (biological evolution), but because language adapted to our brains (cultural evolution).

In all of these culturalist accounts, language change takes centre stage in the evolution of language and biological mechanisms take a back seat. Opponents would argue that language change operates on a superficial level and does not alter the core features of language. In this view, cultural change is minimised and biological conditioning is maximised. For Chomsky, language change simply relates to externalisation of the language faculty. It has no genetic basis or consequence and therefore does not influence the internal faculty of language.

But the reality is that language has two sides. Firstly, it belongs to the natural world, a consequence of genetics, and is therefore a topic of natural science and biology. Lions raw and humans speak. How we learn and process language is determined by genetic constraints. However, language cannot be compared to animal cries, which remain constant from generation to generation. It is malleable and develops through social interaction. In addition to being part of the natural world, language also belongs to the world constructed by humans, and is therefore a topic of social science. It is convenient that both the natural and social sciences adopt the scientific method of research. The task is to systematise findings if we are to include insights from both in our theory of language evolution.

The "gene-culture co-evolution" approach, e.g. Deacon (2010) [20], is an example of how cultural change can be causally interwoven with genetic change. It assumes that the process of cultural evolution plays an active role in the evolution of genes. This view is supported by studies that suggest that many human genes have recently changed as a result of adaptation to new environments created by farming and agriculture [21]. Just as the genes for lactose tolerance appear to be more dominant in typical dairy farming areas of the world, so people's genetic makeup could have changed as a result of a new "linguistic" environment. However, there are objections. Firstly, some would say this model can be perceived as circular. Where did the linguistic environment come from, if not from the brain? Secondly, complex language would have developed too quickly for genetic change to keep up. Thirdly, if language change does play an active role in the evolution of genes, are we to assume that different languages produce different types of brains in humans? As we saw in Chapter 7, it can be demonstrated that different languages create different neural pathways, but this does not involve alterations to an individual's genetic code (see subsequent discussion on epigenetics). Up to now, no matter how different languages appear in their sounds, grammar and vocabulary, any healthy human child can learn any human language without instruction.

Mendívil-Giró [17] takes the more "Chomskyan" view that language change (i.e. cultural evolution) occurs "alongside" the biological evolution of the human language faculty but is not a causal factor of it. He argues that different languages may appear remarkably different on the surface but will always allow their users to perform the same cognitive and communicative functions. In no way does language change produce a qualitative transformation of languages. He notes, "As far as we know, linguistic changes never destroy languages or render them unusable in terms of their usual functions" [17]. Since languages do not vary indefinitely, historical change must be kept in check by the biological constraints of the language faculty. After all, too much diversification would be disadvantageous for communication and ultimately destroy the system.

The scheme outlined in Fig. 10.4, adapted from Mendívil-Giró (2019) [17], attempts to synthesise stages of the biological evolution of the language faculty with the cultural evolution of language. In the upper part of the diagram, FLx and FLy could represent the different language capacities of Homo erectus and Homo sapiens (assuming the former is an ancestor of the latter. FLx could give rise to a specific language Lx_1 which, due to historical language change, could produce other descendant languages Lx_2, Lx_3, etc. All languages of the Lx type would be confined to the biological constraints of FLx. Once evolutionary changes in the architecture of the brain give rise to FLy, all the languages resulting from this new capacity will have common properties different from the previous ones.

This model certainly appeals in its simplicity. While it clearly places biological evolution at the top of the scheme, it does not ignore the process of cultural evolution. Nevertheless, the fact remains that language is an externalised object that has to be learnt and is thus reliant on exposure and social learning. In this respect, Kirby's loop of three adaptive systems (language, learning and genes) is perhaps a more promising model. By introducing the iterated learning paradigm into the scheme,

Fig. 10.4 Mendívil-Giró's
synthesis of the biological
and cultural evolution of
language (adapted from
Mendívil-Giró 2019, Fig. 5)
[17]

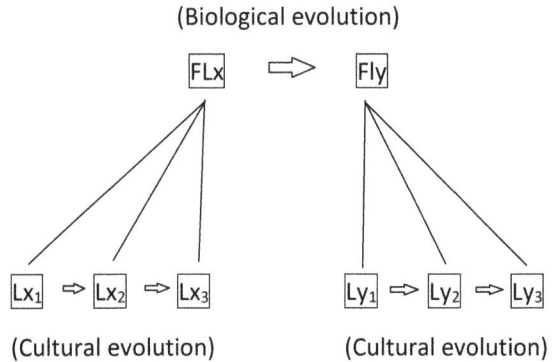

(Biological evolution)

FLx ⇨ Fly

Lx₁ ⇨ Lx₂ ⇨ Lx₃   Ly₁ ⇨ Ly₂ ⇨ Ly₃

(Cultural evolution)   (Cultural evolution)

Kirby provides a framework to explore possible solutions to "Darwin's problem", that is, how such a complex and highly structured system as human language could have emerged through natural selection. While Kirby emphasises the role of cultural processes in shaping language, he acknowledges the influence of genetic biases and biological constraints on the development and structure of language.

From the opposite starting point to that of Kirby, prominent biolinguist Cedric Boeckx clearly places the focus on the biological underpinnings of language. He is particularly interested in the genetic and neural changes that distinguish humans from other primates and make language possible. Nevertheless, he stresses that the biological evolution of the capacity for language depends on the cultural evolution of language and vice versa: on the one hand, linguistic structures arise out of biological structures; on the other hand, culture shapes the evolution of cognition. The result is that language is a "bio-cultural hybrid" [22]. He stresses further that linguistic structure does not emerge in a vacuum—the communicative context matters. According to Boeckx [23], continued success of the biolinguistics enterprise pioneered by Chomsky will depend on recognising that the emergence of many grammatical features is the product of cultural evolution and social transmission. Boeckx believes that recognising the influence of environmental factors (typically resisted in Chomskyan circles) can enrich the biolinguistics enterprise especially in the context of self-organised systems and interaction among agents. He additionally notes that the neo-Darwinian modern synthesis, with its emphasis on adaptation and selection, was "the perfect incubator for evolutionary psychology" but that exclusively selectionist approaches to language evolution are "out of tune with mainstream biology" [23].

As mentioned in the first chapter of this book, the new kind of biolinguistics advocated by Boeckx has moved away from the concepts of theoretical linguistics to focus on data generated by biologists. Yet evolutionary biology today has undergone a further synthesis with the recognition that evolutionary biological dynamics can be radically altered by factors such as ecological interactions and developmental constraints. The current trend is to integrate ecology and developmental biology into evolutionary biology and examine their relationship. The so-called "eco-evo-devo" framework (ecological evolutionary developmental biology) incorporates multiple overlapping biological mechanisms, which are now finding relevance in evolutionary linguistics.

Already Lenneberg [24] pointed out the problem of using natural selection alone to explain language structure and only paying attention to the "biological useful-ness" of certain features. His critical period hypothesis implied a complex inter-action between genetic predispositions and environmental influences at work in the development of language abilities, aligning with the important distinction in genetics between genotype and phenotype. Genotype refers to the set of genes an organism inherits from its parents that can determine traits of that organism. Phenotype refers to the observable traits or characteristics of an organism that result from the interaction between its genotype and its environment. It encompasses all physical, biochem-ical and behavioural features of an organism, such as its appearance, behaviour, metabolism and physiological functions. Not all organisms with the same geno-type look or act the same way because appearance and behaviour are modified by environmental and growing conditions.

One of the questions being discussed within this new framework of evolutionary research is the role played by self-domestication in shaping human cognition and social behaviour. The hypothesised self-taming of humans, resulting from a pref-erence for individuals exhibiting collaborative and social behaviours, is thought to have contributed to our ultra-social phenotype, bringing benefits to the whole of society. According to Boeckx [25], self-domestication changed the way in which humans communicated and shared knowledge, thus shaping our ability to learn and use language. He notes, "If social pressures truly impact grammatical structure ... then the changes in social dynamics brought about by self-domestication must have modified our language capacity" [25]. It is currently Boeckx' aim to pinpoint genetic changes associated with self-domestication traits (such as the loss of aggression and changes in facial morphology) using comparative genetic evidence from animals and archaic humans.

Another area of interest for biolinguists today concerns the role of genes and epige-netics in shaping the development of language-related brain regions and networks. In Chapter 4, we saw that human brain growth was no doubt assisted by the devel-opment of salivary amylase, allowing for the more efficient digestion of starch-rich foods. Studies have shown that a high-starch diet can trigger epigenetic modifica-tions (functionally relevant changes in the genome that alter how genes are expressed, without mutation of the nucleotide sequence) that can influence the activity of the AMY1 gene, thereby affecting the production of salivary amylase. In a similar way, cultural practices, such as language use and social interaction, can trigger epigenetic modifications that influence gene expression patterns related to neural development, synaptic plasticity and cognitive processes involved in using and acquiring language. Over time, these epigenetic changes can become heritable, potentially leading to the evolution of biological mechanisms that are better adapted to the linguistic environ-ment. The result is a feedback loop between cultural practices and biological mecha-nisms. As cultural practices shape the development and functioning of the brain, these biological mechanisms, in turn, influence individuals' ability to engage in cultural activities such as language learning and communication. This bidirectional relationship between culture and biology allows for the co-evolution of linguistic behaviours and underlying neural processes.

To summarise, the modern view is that cultural practices can exert profound effects on the biological mechanisms that support language learning and processing. By shaping genetic, epigenetic and neuroplastic processes, cultural practices influence the development and evolution of language-related cognitive abilities and neural circuits across generations. Co-evolutionists still debate whether it was primarily genetic evolution that drove the development of human culture, or whether cultural evolution was the driving force behind genetic evolution. Studies of living apes suggest that culture has been at least a minor part of hominin capacities since our LCA with chimpanzees and there is some empirical evidence that suggests that culture rather than genes provides more scope for the evolution of large-scale human prosociality. Improved genomic methods in the future will no doubt improve our understanding of the extent of culture-led gene-culture co-evolution in hominin evolution [21].

I believe I have made a compelling case for language being a complex adaptive system shaped by multiple processes occurring over widely diverse timescales, including the accumulation of genetic pre-adaptations, the co-evolution of these pre-adaptations, the emergence of linguistic structure and the selection of motor capacities and learning biases that have enabled humans to excel in language acquisition and use. In addition, language change arising in response to various internal and external pressures plays a crucial role in this complex adaptive system. At the beginning of the twenty-first century, theories of language evolution still grappled with the tension between the internalist (biological) and externalist (cultural) perspectives. However, since then, the interaction between the biological evolution of the mental capacity for language and the ongoing cultural evolution of language has become an exciting area of research. Computer modelling has proven invaluable in such efforts, helping to elucidate the vast interconnectedness of genes, culture and environment, and shedding light on the intricate dynamics that drive the evolution and development of language.

I would like to finish by drawing attention to an exciting new tool for comparing and linking theories and models of language evolution. The Causal Hypotheses in Evolutionary Linguistics Database (CHIELD) [26] is a database of hypotheses, expressed as causal graphs, which allows researchers to explore connections between theories with the goal of "building chains of models that feed into each other". Its creator, evolutionary linguist Dr. Sean Roberts from Cardiff University, notes that we are amassing more and more data on language and more and more data to link it with other domains, but there is little discussion about how to integrate results from different methods. Instead of selecting a single methodology to test a hypothesis, we should use as many tests as possible to achieve a space of results that suggests how robust the hypothesis is [27]. Roberts firmly believes that CHIELD demonstrates that the field of evolutionary linguistics is more interconnected than previously imagined, and that there are many more links between theories waiting to be uncovered [26].

Finally, we need to keep in mind that language is a communication system. Communication, from the Latin *communicare*, meaning "to share", is a means of social cooperation, which is ultimately a biological means of survival. It is a tool used not only by humans and animals but also by plants and microorganisms. Humans have a highly symbolic and structured system of communication that ingeniously uses a finite set of arbitrary signs to convey endless thoughts and messages. Due to its complexity and high level of abstraction, it relies heavily on cultural transmission, which in turn requires advanced vocal learning capacities, as well as a myriad of underlying general learning and social cognitive skills. Clearly, language has a biological foundation upon which language acquisition and language use were able to build, but ultimately language exists and survives due to its shared use.

To conclude, evolutionary linguistics has established itself as one of the fastest-growing subdisciplines in linguistics and is attracting ever more attention from scholars outside of linguistics. In the words of evolutionary biologist and cognitive scientist W. T. Fitch, the evolution of language is "an issue absolutely central to understanding our own species: how we rose from being an ecologically peripheral African ape, a few million years ago, to the most influential (and dangerous) species alive on our planet today" [1]. Scholars have of course always been interested in the origins of language, but it is only recently that researchers have been able to draw from a wealth of data and methodological approaches from a multitude of disciplines, which makes the field both fascinating and challenging. With continued interdisciplinary collaboration among theoretical linguists, evolutionary biologists, anthropologists, anatomists, neuroscientists, geneticists, cognitive psychologists, palaeontologists, archaeologists, social scientists and mathematicians, not to mention the potential contributions of AI, a much deeper understanding of this age-old problem could soon be within our grasp.

References

1. Fitch, W. T. (2017). Empirical approaches to the study of language evolution. *Psychonomic Bulletin & Review, 24*, 3–33. https://doi.org/10.3758/s13423-017-1236-5
2. Christiansen, M., & Chater, N. (2008). Language as shaped by the brain. *Behavioral and Brain Sciences, 31*, 489–558. https://doi.org/10.1017/S0140525X08004998
3. Fitch, W. T. (2018). The biology and evolution of speech: A comparative analysis. *Annual Review of Linguistics, 4*(1), 255–279. https://doi.org/10.1146/annurev-linguistics-011817-045748
4. Fisher, S. E. (2017). Evolution of language: Lessons from the genome. *Psychonomic Bulletin & Review, 24*(1), 34–40. https://doi.org/10.3758/s13423-016-1112-8
5. Miyagawa, S. (2017). *Integration hypothesis: A parallel model of language development in evolution.* In S. Watanabe, M. Hofman, & T. Shimizu (Eds.), *Evolution of the brain, cognition, and emotion in vertebrates.* Brain Science. Springer.
6. Hauser, M. D., Chomsky, N., & Fitch, W. T. (2002). The faculty of language: What is it, who has it, and how did it evolve? *Science, 298*, 1569–1579. https://doi.org/10.1126/science.298.5598.1569
7. Berwick, R., & Chomsky, N. (2011). The biolinguistic program: The current state of its development. In A. M. Di Sciullo & C. Boeckx (Eds.), *The biolinguistic enterprise. New perspectives*

on the evolution and nature of the human language faculty (pp. 19–41). Oxford University Press.

8. Berwick, R. C., & Chomsky, N. (2016). *Why only us: Language and evolution.* MIT Press.
9. Jackendoff, R. (2002). *Foundations of language: Brain, meaning, grammar, evolution.* Oxford University Press.
10. Fujita, H., & Fujita, K. (2021). Human language evolution: A view from theoretical linguistics on how syntax and the lexicon first came into being. *Primates, 63*(5), 403–415. https://doi.org/10.1007/s10329-021-00891-0
11. Jackendoff, R. (2003). Précis of foundations of language: Brain, meaning, grammar, evolution. *Behavioral and Brain Sciences, 26,* 651–707. https://doi.org/10.1017/S0140525X03000153
12. Lust, B. (2006). *Child language: Acquisition and growth.* Cambridge University Press.
13. Progovac, L. (2016). A gradualist scenario for language evolution: Precise linguistic reconstruction of early human (and Neandertal) grammars. *Frontiers in Psychology, 7,* Article 1714. https://doi.org/10.3389/fpsyg.2016.01714
14. Evans, N., & Levinson, S. (2009). The myth of language universals: Language diversity and its importance for cognitive science. *Behavioral and Brain Sciences, 32*(5), 429–448. https://doi.org/10.1017/S0140525X0999094X
15. Smith, K. (2006). Cultural evolution of language. In E-CÂÂ. K. Brown (Ed.), *Encyclopedia of language & linguistics* (2nd ed., pp. 315–322). Elsevier.
16. Heine, B., & Kuteva, T. (2007). *The genesis of grammar.* Oxford University Press.
17. Mendívil-Giró, J.-L. (2019). Did language evolve through language change? On language change, language evolution and grammaticalization theory. *Glossa: A Journal of General Linguistics, 4*(1), 124. 1–30. https://doi.org/10.5334/gigl.895
18. Bybee and Hopper (2001), as noted in Newmeyer, F. J. (2003). *Grammar is grammar and usage is usage.* Project MUSE, John Hopkins University.
19. Bybee, J. (2003), as noted in Hall, J. K., Cheng, A., & Carlson, M. T. (2006). Reconceptualizing multicompetence as a theory of language knowledge. *Applied Linguistics, 27*(2), 220–240. https://doi.org/10.1093/APPLIN/AML013
20. Deacon, T. (2010). A role for relaxed selection in the evolution of the language capacity. *Proceedings of the National Academy of Sciences, 107*(2), 9000–9006. https://doi.org/10.1073/pnas.0914624107
21. Richerson, P. J., Boyd, R., & Henrich, J. (2010). Gene-culture coevolution in the age of genomics. *Proceedings of the National Academy of Sciences, 107*(2), 8985–8992. https://doi.org/10.1073/pnas.0914631107
22. Boeckx, C. (2021). Language Evolution. Pluralism instead of Minimalism. ABRALIN AO VIVO 2023
23. Boeckx, C. (2014). What can an extended synthesis do for Biolinguistics? In M. Pina & N. Gontier (Eds.), *The evolution of social communication in primates* (Vol. 1). Springer.
24. Lenneberg, E. H. (1967). *Biological foundations of language.* Wiley.
25. Boeckx, C. (2021). Reflections on language evolution: From minimalism to pluralism. In M. Dingemanse & N. J. Enfield (Eds.), *Conceptual foundations of language science 6.* Language Science Press.
26. Roberts, S. G., et al. (2020). CHIELD: The causal hypotheses in evolutionary database. *Journal of Language Evolution, 5*(1), 1–20. https://doi.org/10.1093/jole/lzaa001
27. Roberts, S. G. (2018). Robust, causal, and incremental approaches to investigating linguistic adaptation. *Frontiers in Psychology, 9.* https://doi.org/10.3389/fpsyg.2018.00166

Glossary

Acoustic processing Initial stage in *speech perception* whereby acoustic features such as *pitch*, intensity and duration are detected and used to break down the speech signal into meaningful elements.

Adaptation (i) Evolutionary process of change that fits organisms or species to their environment, involving *natural selection.*

(ii) Trait in a species that has evolved over generations, through *natural selection*, in response to a new environment.

Adaptationism View that language evolved slowly through genetic *adaptation* and *natural selection.*

Adaptationists Those who believe human language is an evolved *adaptation*, developing slowly with incremental changes taking hold through fitness advantages.

Affix *Morpheme* that cannot stand alone and must be attached to a *word root,* either at the beginning (= prefix) or at the end (= suffix). Can indicate grammatical relationships such as number and tense (e.g. clock-**s**, walk-**ed**) or modify the meaning of a word (e.g. **un-**happy).

Agent (i) One of the *semantic roles* of a noun phrase within a sentence. The thing or person that does an action.

(ii) Language user in a model simulating language behaviour.

Agent-based models Computational models that simulate the actions and interactions of autonomous *agents* (e.g. language users) in order to understand the behaviour of a system (e.g. language).

Allophones Sound variants of a single *phoneme* that do not produce a change in word meaning. Some allophones ease awkward articulations, e.g. aspirated t in *top* becomes unaspirated t in *stop.* Other allophones are dialect-dependent, e.g. intervocalic t in *water*, pronounced [t] in Standard British English, becomes flapped t (= [ɾ]) in American English.

Analogous trait Similar feature that evolved independently among genetically far removed species.

J. Dornbierer-Stuart, *The Origins of Language*,
https://doi.org/10.1007/978-3-031-54938-0

Analysis Breaking down a whole into its component parts. The "analytic model of language evolution" suggests humans started with *holistic* vocalisations and slowly isolated individual chunks and associated these with meaning.

Animal cognition Mental capacities of non-human animals. Central source of information relevant to the evolution of human *cognition* and, thus, language.

Aphasia Inability to produce fluent speech (*Broca's aphasia*) or meaningful speech (*Wernicke's aphasia*) because of damage to specific brain regions.

Arbitrariness Absence of any natural connection between a word's meaning and its sound or form.

Archaeology Study of human history and pre-history via material *culture*.

Articulation (i) Process of producing a speech sound. Also refers to the resulting speech sound.

(ii) In *speech processing*, the final stage in *speech production* involving the physical production of an acoustic speech signal.

Attention Cognitive mechanism that enables us to highlight relevant features of input data. It allows information to be prioritised for *speech processing*.

Australopithecus Genus of early *hominins* that emerged during the *Pliocene* and from which *Homo* evolved.

Babbling Stage in children's *language acquisition* during which an infant appears to be experimenting with producing sounds. Also occurs in some animals and songbirds.

Bare phonology Patterning of sounds in *holistic* messages or tunes that refer to whole ideas or situations and which cannot be broken down into smaller meaningful parts.

Behaviourism View that all behaviours, including language, are a set of habits acquired by *imitation* and reinforcement. It ignores the mental processes that underlie behaviour and focuses solely on output.

Behaviourists In linguistics, those who put a strong emphasis on the environment in the development of human language. They argue language is a set of habits acquired by *imitation* and reinforcement.

Biolinguistics Discipline that seeks to provide a biological framework to understand the fundamentals and evolution of language.

Bipedalism Ability to walk on two feet. Humans as well as some other groups of modern species (e.g. birds, kangaroos) are habitual bipeds whose normal method of locomotion is two-legged.

Broad Language Faculty (FLB) Set of all mechanisms, *shared* or *derived*, involved in acquiring, perceiving and producing language.

Broca's aphasia Impairment or loss of the ability, following damage to *Broca's area* of the brain, to produce fluent speech. Comprehension generally remains intact.

Broca's area Cluster of interconnected areas located in the inferior frontal gyrus of the frontal lobe of the brain. Associated with vocal control and *syntax*.

Case system System of inflections occurring in nouns, pronouns, adjectives and articles which indicate causal relations in a sentence. The most frequently encountered cases are nominative, accusative, dative and genitive.

Categorical perception Innate ability to perceive categories along a continuum (e.g. colours along the colour spectrum or *phonemes* along the sound spectrum). In human language, boundaries between categories are determined largely by convention.

Categorisation Ability to group objects or elements that are alike (including syntactic constituents).

Causality Cause-effect relations. Causal cognition in humans is central to explaining sentence structure. Sentences express events, which are based on who does what to whom.

Cerebral connectivity Neural connections between different areas of the brain. Modulated by learning and the environment.

Child-directed speech Special way of talking to an infant, characterised by a slower pace, shorter utterances, exaggerated pitch modifications and special vocabulary.

Cognition All functions and processes of the mind, including perception, *attention*, thought, imagination, intelligence, *memory*, reasoning, problem-solving, decision-making and language.

Cognitive linguistics Movement in linguistics that grew as a reaction to *generative grammar*. A fundamental tenet of cognitive linguistics is that the cognitive capacities that speakers and hearers employ in using language are domain-general: they underpin not only language, but also other areas of *cognition*.

Cognitive neuroscience Scientific study relating neural activity in the brain to cognitive functions such as *memory*, learning, consciousness and language.

Cognitive science Scientific study of all functions and processes of the mind, with input from linguistics, psychology, neuroscience, philosophy, computer science and anthropology.

Combinatorial phonology Combining of meaningless elements (speech sounds) to form meaningful elements (*morphemes* and words).

Combinatorial structure Feature of language whereby smaller elements, such as sounds or words, are combined to form larger elements, such as words or sentences.

Communicative competence A language user's unconscious knowledge of the whole language system, including how to use it appropriately.

Comparative approach (i) Comparative study of structural features across different languages to highlight commonalities or diversity.

(ii) Comparative study of cognitive abilities across species with a view to understanding the evolutionary path to human language.

Comparative historical linguistics Comparative study of languages for evidence of ancestral *proto-languages*.

Compounding Combining existing words to form new words, e.g. *cry-baby*.

Conceptualisation (i) In *cognitive science*, the process of making mental representations of objects and concepts and mapping relationships between them.

(ii) In *speech processing*, the first stage in *speech production* involving selection of the concepts to be expressed and identification of the relationships between them.

Consonant Speech sound produced with complete or partial closure of the *vocal tract*, e.g. [p], [d] and [m]. Usually used at the beginning (onset) or end (coda) of a *syllable*.

Content words Lexical words, i.e. nouns, verbs and adjectives, carrying *referential meaning*. The "bricks" in a wall.

Continuity View that language has evolved from precursors in the animal world.

Convergence Process in *sociolinguistics* whereby speakers adapt their language towards that of the interlocuter to achieve social acceptance.

Convergent evolution Independent development of similar features in species with different ancestral origins.

Creole *Pidgin* that has become a group's first language and therefore undergoes dramatic expansion.

Critical period In developmental biology, a maturational period in the life of an organism when the nervous system is especially sensitive to certain environmental stimuli in order to develop a skill that is indispensable for the organism's survival. In *language acquisition*, a period during which language must be acquired in order to insure the development of native competence.

Cultural diffusion Process by which *culture*, including language, spreads gradually across a community through social interaction.

Cultural evolution *Language change* through social interaction and *cultural transmission* rather than genetic inheritance.

Cultural selection theory Theory that models cultural change on Darwinian *natural selection*. An extension of *memetics*.

Cultural transmission Process by which *culture*, including language, is passed on from one generation to the next through *imitation* and learning.

Culturalism View that language is primarily culturally constructed and socially transmitted. In its strongest form, it can account for the genesis of language.

Culturalists Those who believe human language is primarily culturally constructed and socially transmitted.

Culture (i) Immaterial culture. Social behaviour and knowledge acquired through *imitation* and learning. Includes expressive forms like art, music, dance, ritual and religion; technologies like tool usage, cooking, shelter and clothing, and "virtual realities" such as concepts, beliefs, institutions and language.

   (ii) Material culture. Physical expressions of culture, such as tools, artefacts, architecture and cultural landscapes.

Derivation Process of forming a new word by adding a prefix or suffix to an existing word.

Derived trait Feature evolving uniquely in a species, e.g. the large brain in humans.

Diachronic approach Approach in linguistics that considers the development of language over time.

Diachronic linguistics See *Historical linguistics*.

Discontinuity Principle that a biological capacity in a species can evolve from scratch in that species.

Discreteness Feature of language that uses a limited set of sounds far enough apart on the continuum of possible sounds so that each sound remains distinctive.

Displacement Feature of language that allows us to communicate ideas that are remote in time and space.

Divergence (i) See *Divergent evolution.*

(ii) Process in *sociolinguistics* whereby speakers distance themselves from the interlocuter by steering their language away from that of the interlocuter.

Divergent evolution Development of differences within a species that can lead to *speciation.*

DNA A molecule present in the cells of all living organisms that carries genetic information for the development and functioning of an organism.

Duality of patterning Feature of language whereby meaningless elements (speech sounds) are combined into meaningful elements (words), which are combined further into phrases and sentences.

Dyslexia Disorder characterised by a difficulty in reading and writing. Can be due to brain damage, inherited factors or unknown causes.

E-language Language as externalised and transmitted by the community. The tangible, observable expression of language that can be recorded and studied by scientists and linguists.

Eco-evo-devo Field of biology that studies the interactions between ecology, developmental biology and evolutionary biology.

Electroencephalography (EEG) Method used for recording the electrical activity of the brain. Electrodes placed on the scalp are linked to an electroencephalograph, which produces visible brainwave patterns.

Embodiment Use of the motor system to link mental concepts with sensory experience.

Epigenetics Study of stable changes in the *genome* that affect gene expression without altering the underlying *DNA* sequence.

Event-related potentials (ERPs) Changes in the brain's electrical activity in response to specific stimuli. In the context of *language processing*, the so-called N100 (peaking at 100 milliseconds) has been linked to initial *acoustic processing*, the ELAN (peaking at 200 milliseconds) to syntactic structure building and the N400 (peaking at 400 milliseconds) to semantic processing and *thematic relations* assignment.

Evolutionary linguistics Discipline that tries to explain the emergence and subsequent development of language in humans. Deals with the biological evolution of the capacity for language as well as the cultural development of language. Being a scientific discipline that relies on empirical data, its main challenge is that there are no direct physical traces of early human language.

Externalisation Expressing outwardly what is originally internal. For example, an intention can be externalised as an action, an emotion as a facial gesture or a thought as speech.

Faculty of language (FL) Uniquely human cognitive system proposed by Noam Chomsky that supports the acquisition and use of language.

FOXP2 gene Highly conserved *gene* in mammals. The version in humans differs from that in non-human primates by the substitution of two amino acids. It is theorised that these changes were essential for vocal control and thus language.

Function words Grammatical words, i.e. articles, prepositions, pronouns and conjunctions, that fill the gaps between *content words*. The "cement" between the bricks.

Functionalist approach Method in linguistics that seeks to explain language structure in relation to what language is used for.

Gene One of many stretches of *DNA* in a chromosome that contributes to the specification of some trait of an organism.

Gene-culture co-evolution Theory that proposes *culture* plays an active role in the evolution of *genes*.

Gene flow Changes in the *gene pool* brought about when individuals migrate from one population to another.

Gene pool Set of all *genes* in a population.

Generative grammar (i) Linguistic theory that views language as an innate syntactic structure, which appeared suddenly in humans and is autonomous from other cognitive systems.

(ii) Noam Chomsky's proposed system of *syntactic rules* that generate all the possible grammatical sentences of a language and from which phonological and semantic structures are subsequently derived.

Genetic drift Changes in the composition of a population's *gene pool* that are brought about by random chance rather than *natural selection.*

Genetic mutation Changes in the *DNA* sequence of an organism. Means by which nature creates variation in species.

Genome Complete set of genetic information in an organism. Provides all of the information an organism requires to function. Stored in chromosomes (long molecules of *DNA*) found in the nucleus of most living cells.

Genotype Set of *genes* an organism inherits from its parents that can determine traits of that organism.

Gradualists Those who believe that the development of language was a step-wise process, with each stage building upon and modifying the previous stage. Gradualism can refer to both the biological and *cultural evolution* of language.

Grammar (i) Prescriptive set of rules setting forth the current standard version of a language. Mainly deals with the language's *syntax* and *morphology.*

(ii) In linguistics, the subconscious set of knowledge speakers have about the categories of language (phonological, morphological and syntactic) and the rules by which they interact. Sometimes refers just to the rules of *syntax.*

Grammaticalisation Process of *language change* whereby lexical items (nouns and verbs) develop over time into grammatical items such as auxiliaries, case markers, inflections and prepositions. Believed by some to be a key mechanism in the origin of *grammar.*

Great Rift Valley Term often used for the East African Rift System that developed around the onset of the *Miocene*. Shifts to more arid, open conditions produced major steps in the evolution of *hominins.*

Great Vowel Shift A gradual shift in the pronunciation of all Middle English long vowels (/aː ɛː eː iː ɔː oː uː/).

Hierarchical structure Organising principle in language whereby smaller elements (e.g. words) are embedded into larger elements (e.g. phrases), which are in turn embedded into yet larger elements (e.g. clauses and sentences).

Historical linguistics Study of *language change* over time and the development of theories to explain the underlying processes.

Holistic Concerned with wholes rather than separation into parts. Holistic vocalisations are vocalisations that refer to whole ideas or situations and cannot be broken down into smaller meaningful parts.

Holocene Current geological epoch starting around 12,000 years ago. The Holocene coincides with the rapid growth and proliferation of humans and the transition to urban living.

Holophrase Single word expressing the whole idea in a phrase or sentence, e.g. *Down!* to mean "I want to get down".

Hominidae (hominids) Taxonomic family commonly called the great apes, including the genera Pongo (orangutans), Gorilla, Pan (chimpanzees) and *Homo*.

Hominini Taxonomic branch of the great apes that includes the genera Pan (chimpanzees) and *Homo* (humans).

Hominina (hominins) Taxonomic branch of the *Hominini* that includes *Australopithecus* and *Homo* (from which modern humans emerged).

Homo Genus that emerged from *Australopithecus* encompassing several extinct species of humans (e.g. Homo habilis, Homo erectus) plus the extant species Homo sapiens (modern humans).

Homologous trait Feature shared with another species through a common ancestor.

Horizontal transmission Transmission of language across a community through social interaction.

Icon Type of sign that resembles its *referent* through its form. Words may resemble their referents through how they sound (e.g. *splash*) or how they are structured (e.g. *toothbrush*).

Iconicity Existence of a natural connection between a sign's meaning and its sound or form.

I-language Intrinsic linguistic knowledge held in the mind of a speaker. Can be observed indirectly through a speaker's "intuitions" about what constitutes a possible sentence in their language.

Imitation Process whereby an individual observes and replicates another's behaviour, allowing the transfer of behaviour without the need for genetic inheritance.

Internalisation Converting what is perceived externally into a feature of one's interior (cognitive) landscape. For example, the rules of *grammar* of one's native language are internalised as linguistic knowledge.

Intonation *Pitch* contour of speech that can serve to indicate the grammatical function of an utterance (e.g. statement vs. question), as well as reveal the attitudes and emotions of a speaker.

Intonational phrase Chunk of speech typically carried out in one breath and centred around a prominent word, e.g. / the girls in the house were singing lullabies /.

Intraspeaker variation Individuals' style-shifting, causing language variation and change over time.

Iterated learning Process whereby the learnt output of one generation becomes the input from which the next generation learns. Emphasises the significance of *cultural transmission* in *language evolution*.

Language acquisition Process by which humans learn to use language. Requires exposure to the system and the use of innate capacities such as *categorisation*, *imitation*, *memory* and problem-solving.

Language Acquisition Device (LAD) Innate mental capacity proposed by Noam Chomsky for acquiring language.

Language change Modification of language forms over time. The subject of interest in several subfields of linguistics: *historical linguistics*, *sociolinguistics* and *evolutionary linguistics*.

Language evolution Emergence and subsequent development of language in humans. A co-evolved continuum involving biological, cultural and social forces.

Language faculty Biological capacity of humans to learn and use language.

Language processing Mental processes involved in understanding, producing and acquiring any form of language, including spoken, written and *sign language*.

Language universals Set of structural features shared by all human languages, pioneered by Joseph Greenberg in the 1960s and 70s. Mostly concerned with the categories of *syntax* (words, *morphemes*) and how these are ordered in phrases and sentences.

Language use Use of language in concrete situations for communication. Accounts for considerable variation of language over time.

Larynx Voice box housing the vocal cords, which manipulate the *pitch* and volume of the voice.

Last common ancestor (LCA) In biology, the most recent individual from which two species (e.g. chimps and humans) are descended.

Learning biases Innate preferences and tendencies that individuals have when acquiring a language. For example, language learners are sensitive to regularities in language, which helps them identify linguistic patterns and structures.

Lemma (i) In *morphology*, the canonical or base form of a *lexeme*. Similar to a headword in a dictionary. It is the abstract form underlying any inflected form, e.g. *smiles*, *smiled*, *smiling*, etc., are all *lexemes* of the lemma SMILE.

  (ii) In *psycholinguistics*, the conceptual form of a word selected for utterance. It carries meaning but has not yet undergone morpho-phonological *encoding*.

Lexeme (i) In *morphology*, the smallest sequence of sounds that can stand alone with objective meaning and be moved to different places (grammatical slots) in sentences while staying intact, e.g. *smile*, *wheelbarrows*.

  (ii) In *psycholinguistics*, a *lemma* that has undergone morpho-phonological *encoding*.

Lexical borrowing Importing words into a language from other languages.

Lexical diffusion Gradual spread of a particular sound change to all similar-looking words in the vocabulary of a language.

Lexical meaning See *Referential meaning*.

Lexicon Vocabulary of a language.

Linear sequencing Organising principle in language whereby elements (syntactic constituents) are ordered sequentially (one after the other) rather than *hierarchically*.

Linguistic competence A language user's unconscious knowledge of *grammar*.

Linguistic decoding Process in *speech perception* whereby grammatical information is extracted from a stream of speech in order to arrive at the intended message conveyed by the speaker.

Linguistic encoding Process in *speech production* whereby thoughts are transformed into a linguistic form, containing syntactic, morphological and phonological information, and finally given phonetic structure, ready for *articulation*.

Linguistic reconstruction Method used to reconstruct earlier forms of a language by using evidence from correspondences in later languages.

Linguistic variation Different ways of saying the same thing. Variation occurs in all levels of *grammar* and can be attributed to physiological, psychological and sociological factors.

Long-term memory Cognitive system that holds information indefinitely through synapse enlargement and stabilisation. Important for word recognition (in *speech perception*) and word retrieval (in *speech production*).

Manner of articulation Way in which air escapes from the *vocal tract* while producing a speech sound. Can involve substantial obstruction, as with stops and fricatives, or little or no obstruction, as with nasals, approximants and *vowels*.

Memetics Theory of information transfer proposed by Richard Dawkins, according to which memes, much like *genes* in biology, are units of information passed through generations of *culture*.

Memory Cognitive system in which data or information is perceived, stored and retrieved when needed.

Mental lexicon A person's "mental dictionary". Set of knowledge a speaker has about words and their meanings.

Merge A structure-building operation in the *Minimalist Program* that allows us to tag words onto words *recursively* to create an infinite number of expressions.

Minimalist Program Model of human language proposed by Noam Chomsky that sees language as a human-specific computational system (known as *Merge*) that allows us to create an infinite number of novel sentences.

Miocene Geological epoch extending from about 23 to 5 million years ago. The late Miocene saw the emergence of *Hominina (hominins)*.

Mirror neurons A type of neuron that is thought to cause a perceived action to be simulated in the perceiver's brain. Considered to form the basis for *imitation* and social learning.

Modality Channels involved in different forms of language. Speech involves the vocal-auditory modality and *sign language* involves the gestural-visual modality.

Modularity Idea that a system is composed of independent modules that interact and function as a whole. In linguistics, the term is used in two ways, either to refer to the *faculty of language* as a distinct module of the mind, or else to refer to the fact that the human language capacity is composed of various modules that handle different aspects of language.

Morpheme Smallest meaningful part of a word, either a *word root* or an *affix*. A *word root* (e.g. CLOCK) generally refers to something existing in the outside world and may be unadjoined, whereas an *affix* (e.g. -S) only carries grammatical meaning and cannot stand alone.

Morphology (i) Study of words, their meaning and how they are structured.

(ii) Component of *grammar* that deals with words, their meaning and how they are structured.

Motherese See *Child-directed speech.*

Multimodal Involving a variety of modes (e.g. vocal, gestural) or *modalities*— language uses the vocal-auditory modality for speech and the gestural-visual modality for *sign language.*

Mutation See *Genetic mutation.*

Naming insight The moment when a child realises that things have labels. Some believe the naming insight was a crucial moment in the evolution of human language.

Narrow Language Faculty (FLN) Proposed "core" of language which is unique to humans and specific to language. The computational mechanism for *recursion.*

Nativism View that we are born with a certain knowledge of how language works.

Nativists Those who believe we are born with a certain knowledge of how language works.

Natural selection Key mechanism of evolution whereby organisms that are better adapted to their environment tend to produce more offspring and transmit more of their genetic characteristics to succeeding generations than those that are less well adapted.

Neurodevelopment Formation of the neurological connections and pathways in the brain that mediate a highly complex set of behaviours, including language.

Neurolinguistics Discipline that deals with the relationship between language and the structure and functioning of the brain, especially the neural mechanisms that control the comprehension, production and acquisition of language.

Neuroplasticity Ability of neural networks in the brain to respond and adjust to environmental influences, which has implications for learning new skills. The developing brain of a child exhibits a higher degree of plasticity than the adult brain.

Nicaraguan Sign Language (NSL) *Sign language* which developed spontaneously among deaf children in a school in Nicaragua in the 1980s.

Niche construction Process in biology whereby an organism alters its local environment, thus interfering with *natural selection* and changing *gene* frequencies in populations.

Object (O) Grammatical argument that is not the *subject*. Similar but not identical to *patient*. For example, in the sentence *The girl was bitten by the dog*, "dog" is the object, but "girl" is the *patient*.

Onomatopoeia Feature of language whereby a word representing an object or event imitates the sound associated with the object or event. Considered by some to have set the foundation for using speech sounds to refer to concepts.

Ontogeny Development of an individual organism or trait of an organism from inception to maturity. Can refer to the development of language in children.

Paleo-DNA *DNA* recovered from archaic humans.

Palaeontology Study, through fossil remains, of life prior to the *Holocene*.

Panina Taxonomic branch of the *Hominini* from which chimpanzees and bonobos emerged.

Parallel Architecture Model of human language that sees *phonology, syntax* and *semantics* as autonomous generative systems linked by interface components.

Patient One of the *semantic roles* of a noun phrase within a sentence. The thing or person that undergoes an action.

Phenotype Observable traits or characteristics of an organism that result from the interaction between its *genotype* and its environment.

Phoneme Abstract unit of sound that can distinguish one word from another in a particular language, e.g. /p/ and /t/ are phonemes in English since "pin" has a different meaning to "tin".

Phonetics Study of the physical sounds of language and how they are articulated.

Phonological phrase A constituent in the phonological hierarchy that is between *phonological word* and *intonational phrase*. Each phonological phrase contains a prominent word, e.g. [the girls] [in the house] [were singing] [lullabies].

Phonological rules Rules underlying how we pronounce *phonemes* in different phonological environments, either within words (e.g. /n/ in *thorn* assimilates to /m/ in *Thornberry*) or between words (e.g. /d/ in *mad* assimilates to /g/ in *mad goat*).

Phonological word A constituent in the phonological hierarchy that is between *syllable* and *phonological phrase*. Each phonological word contains a prominent *syllable*, e.g. sing-ing.

Phonology (i) Study of how sound is organised to convey linguistic meaning. Includes studying the elementary units of sound (*phonemes*) and how these combine, as well as the features that accompany speech sounds (*suprasegmentals*).

(ii) Component of *grammar* that deals with how sound is organised to convey meaning.

Phonotactic rules Set of constraints on how sequences of sounds are configured. For example, in English, if a word begins with /s/ and the following sound is a stop, that stop must be *voiceless* (/p/, /t/ or /k/). The rules are highly variable between different languages.

Phrase structure grammar Formal method employed by linguists to break down sentences *hierarchically* into phrases and words using an upside-down tree-like structure.

Phylogeny Evolutionary development and history of a species or trait of a species. Can refer to the evolution of language in humans.

Pidgin Simplified means of communication that develops among people who do not have a language in common.

Pitch Highness or lowness of speech determined by the vibratory frequency of the vocal cords.

Place of articulation Point at which obstruction occurs in the *vocal tract* while producing a speech sound.

Pleistocene Geological epoch, commonly known as the Ice Age, extending from around 2.5 million to 12,000 years ago. The early Pleistocene saw the emergence of *Homo*.

Pliocene Geological epoch extending from around 5 to 2.5 million years ago. The early Pliocene saw the emergence of *Australopithecus*.

Pragmatics (i) Study of how context contributes to the meaning of an utterance.
    (ii) Aspect of meaning that comes from the context of an utterance.

Predicate Grammatical category pertaining to everything in a sentence that is not the *subject*.

Principle of maximum differentiation Principle that states that a change in one part of the sound system will cause the rest of the system to reorganise so as to keep the distinction between different words clear.

Productivity The ability of language to make novel utterances.

Propositional meaning That part of meaning conveyed by a sentence that relates to some state of affairs in the world. Propositional meaning remains the same regardless of when or where the sentence is uttered.

Prosody System of *rhythm*, *stress* and *intonation* in speech. Provides important information beyond literal word meaning.

Proto-Human Hypothetical common ancestor of all the world's languages based on *linguistic reconstruction*.

Proto-language A postulated ancestral language, e.g. Proto-Indo-European, from which a number of attested languages are believed to have descended, forming a language family.

Protolanguage A primitive language-like system posited in language origin theories. A hypothesised intermediate stage in the emergence of language that was present in archaic humans, qualitatively different from non-human primate communication and modern language.

Proto-syntax (i) Earliest form of *syntax* produced by infants, featuring two-word utterances.
    (ii) Hypothesised earliest form of *syntax* produced by early humans, consisting of *linear sequencing*.

Proto-word (i) Very early word-like utterance produced by an infant (e.g. *Eh!* used as a request) before true first words appear.
    (ii) Hypothesised word-like utterance produced by early humans before they had the capacity for full language.

Psycholinguistics Discipline that studies and tests theories of how language is processed and represented in the mind. It uses a number of non-invasive techniques to discover how we produce, comprehend and acquire language.

Recursion The repeated coupling of elements (e.g. words) to create larger elements (e.g. phrases). Thought by some to be responsible for producing a boundless system of communication in humans.

Referent The object or idea that a signal or sign refers to.

Referential meaning Aspect of meaning that describes things existing in the world, irrespective of the present situation. The inherent meaning of a word—not dependent on context. Also known as lexical meaning.

Rhythm Pattern of timing in an utterance caused by the succession of stressed and unstressed *syllables*.

Root See *Word root*.

Root word See *Lemma*.

Saltationism View that language emerged in a single and, in evolutionary terms, sudden step.

Saltationists Those who believe human language evolved in a single genetic *adaptation* some time after our evolutionary split from the apes.

Segment Any discrete unit that can be identified auditorily in a stream of speech, e.g. *phoneme*, *syllable* or word.

Self-domestication Hypothesised self-taming of humans. A preference for individuals with traits such as docility and emotional intelligence would have brought changes in social dynamics that are thought to have modified our language capacity.

Self-organisation Idea that the structure and patterns in language are not genetically pre-determined but arise from multiple pressures acting on language as it emerges and changes in socially interacting populations.

Semantic roles See *Thematic relations*.

Semanticity Capacity of a communication system to use signs to represent events, ideas, actions and objects in the real world.

Semantics (i) Study of meaning and how it is conveyed in a language through sounds, words, *syntax*, *prosody* and context.

(ii) Component of *grammar* that deals with how meaning is conveyed in a language.

Semiotics Study of the use of signs (linguistic and non-linguistic) to communicate meaning. Signs may imitate the objects they refer to (*icons*) or represent them in conventionalised forms (*symbols*).

Sexual selection A mode of *natural selection* in which members of one sex compete for access to members of the opposite sex, leading to some individuals having greater reproductive success than others.

Shared trait See *Homologous trait*.

Sign language Language that uses the visual-gestural *modality* to convey meaning. Sign languages are considered fully fledged natural languages with their own *grammar* and *lexicon*.

Social cognition Mental capacities used in social interaction.

Social prestige In *sociolinguistics*, the high regard gained by using a specific language form or variety within a speech community. Prestige may be accorded to the standard form of a language (overt prestige) as well as non-standard forms (covert prestige).

Society Community of interdependent organisms, usually of the same species. A society can offer its members benefits that would not be possible on an individual basis.

Sociolinguistics Discipline that studies the effect of *society* on language and how it varies between different groups. Differences in usage may be geographically, socially or stylistically motivated.

Sound symbolism Perceptual similarity between the sound of a sign and its *referent*. The sound of a sign may imitate the sound of an object it refers to (*onomatopoeia*) or some other sensory property, such as size or shape.

Speciation Evolutionary process whereby populations evolve to become distinct species. May occur through splitting of lineages (e.g. by migration) or through *genetic mutation* and recombination within a population.

Speech error Unconscious deviation from the intended form of an utterance, otherwise known as a slip of the tongue.

Speech melody See *Intonation.*

Speech perception Process whereby the sounds of language are heard, interpreted and understood. Involves *acoustic processing, linguistic decoding* and semantic interpretation.

Speech processing Sensory, motor and cognitive processes involved in understanding and producing spoken language.

Speech production Process whereby thoughts are translated into coherent speech. Involves *conceptualisation, linguistic encoding* and *articulation.*

Stress Emphasis placed on a syllable in a word (word stress) or at various points in a sentence (sentence stress) through increased loudness and vowel length and a rise in *pitch.* Word stress is usually fixed, whereas sentence stress can be used to highlight salient information or create contrast.

Structuralist approach Method in linguistics that aims to break language down into its elements (sounds, words, sentences) and explain how these are organised and relate to one another. Language is conceived as a self-contained system of interconnected units.

Subject (S) Grammatical argument that controls the verb. Similar but not identical to *agent*. For example, in the sentence *The girl was bitten by the dog*, "girl" is the subject, but "dog" is the *agent*.

Suprasegmentals Features of speech, such as *stress, rhythm* and *intonation*, that are not phonetic *segments* (*vowels* and *consonants*) but accompany speech sounds, *syllables* and larger units of speech.

Syllable A group of speech sounds in a word that are pronounced in a single beat. A constituent in the phonological hierarchy that is between *phoneme* and *phonological word.*

Symbol Type of sign that represents its *referent* in a conventionalised form that must be learnt. For example, the word *table* bears no physical resemblance to the object it refers to.

Symbol grounding Cognitive process by which abstract concepts and *symbols* are connected to real-world sensory experience through engagement of the motor system. Crucial for understanding and using language effectively.

Symbolic culture Non-material *culture*, including concepts, beliefs, institutions and language, normally considered to be constructed uniquely by Homo sapiens.

Symbolic niche Self-constructed environment consisting of symbolic behaviour to which humans adapt.

Symbolism Practice of representing objects and ideas with *symbols*. Symbolism allows us to refer to things out of sight, the past or future, the hypothetical and the possible.

Symptomatic signalling Signals resulting from an internal state or emotion of the sender. Often consists of fixed, instinctual responses to specific stimuli.

Synchronic approach Approach in linguistics that considers language at a single moment, often the present day, to obtain a snapshot of its characteristics.

Syntactic rules Rules underlying how words are ordered (e.g. SVO) and constrained (e.g. by agreement) in sentences.

Syntax (i) Study of the rules underlying how we combine words into phrases and sentences.

(ii) Component of *grammar* that deals with the rules underlying how we combine words into phrases and sentences.

Synthesis Putting pieces together as a whole. The "synthetic model of language evolution" assumes words evolved first, followed by syntactic operations for combining them.

Telegraphic speech Two-word utterances representing simple *syntax*, e.g. *Get down* to mean "I want to get down".

Thematic relations The various semantic roles that a noun phrase can play within a sentence (e.g. *agent*, *patient*). At the interface of *syntax* and *semantics*.

Theory of least effort Theory that suggests the sounds in words change to facilitate *articulation*.

Theory of mind Ability to judge other people's mental states, intentions and communicative goals. Especially implicated in pragmatic phenomena such as indirect replies, irony and humour.

Tone Variation in *pitch* that can change the meaning of a word, as occurs in Chinese.

Universal Grammar (UG) Noam Chomsky's hypothesised human-specific cognitive structure that generates grammatical sentences. Set of principles underlying all languages that are hardwired into the human brain.

Universals See *Language universals.*

Vertical transmission Transmission of language from parents to offspring through *imitation* and learning.

Vocal imitation Mechanism whereby an individual observes and replicates another's vocal behaviour. Sensory experience is internalised and used to shape vocal outputs.

Vocal learning Learning of language or birdsong through social interaction using innate mechanisms such as *imitation, memory* and general learning strategies.

Vocal production Ability to articulate speech sounds.

Vocal tract Passageway used in the production of speech, above the *larynx* and including the oral, nasal and pharyngeal cavities.

Voiced Describes a speech sound produced with vibration of the vocal cords.

Voiceless Describes a speech sound produced without vibration of the vocal cords.

Vowel Speech sound produced without obstruction of the *vocal tract*, e.g. [æ] as in *pat* and [ɑ:] as in *father*. A vowel forms the nucleus of a *syllable*.

Wernicke's aphasia Impairment or loss of the ability, following damage to *Wernicke's area* of the brain, to comprehend and produce meaningful speech. Patients may demonstrate fluent speech with intact *syntax*.

Wernicke's area Cluster of interconnected areas located in the superior temporal gyrus of the temporal lobe of the brain. Associated with speech comprehension.

Word order Linear order of syntactic constituents in a phrase or clause. Can refer to the position of modifiers in a noun phrase, the position of adverbials in a sentence or the relative order of *subject*, verb and *object* in a clause. About half of the world's languages deploy subject-object-verb (SOV) word order.

Word root Lexical core of a word (e.g. RUN in *running*) which can function as a word or to which a prefix or suffix may attach. A word root carries *lexical meaning* rather than grammatical meaning.

Working memory Dynamic cognitive system that holds information temporarily through the persistent firing of neurons. Important for reasoning and decision-making as well as *speech processing*.

Bibliography

1. Aitchison, J. (1991). *Language change: Progress or decay?* (2nd ed.). Cambridge University Press.
2. Aitchison, J. (2000). *The seeds of speech.* Cambridge University Press.
3. Arbib, M. A. (2005). From monkey-like action recognition to human language: An evolutionary framework for neurolinguistics. *Behavioral and Brain Sciences, 28*, 105–167. https://doi.org/10.1017/s0140525x05000038
4. Arbib, M. A., Liebal, K., & Pika, S. (2008). Primate vocalization, gesture, and the evolution of human language. *Current Anthropology, 49*(6), 1053–1076. https://doi.org/10.1086/593015
5. Baron-Cohen, S., Tager-Flusberg, H., & Cohen, D. J. (2000). *Understanding other minds: Perspectives from developmental cognitive neuroscience* (2nd ed.). Oxford University Press.
6. Beecher, M. D. (2021). Why are no animal communication systems simple languages? *Frontiers in Psychology, 12*, Art 602635. https://doi.org/10.3389/fpsyg.2021.602635
7. Bell, A. (1984). Language style as audience design. *Language in Society, 13*, 145–204. https://doi.org/10.1017/S004740450001037X
8. Bergelson, E., & Aslin, R. (2017). Nature and origins of the lexicon in 6-mo-olds. *Proceedings of the National Academy of Sciences, 114*(49), 12916–12921. https://doi.org/10.1073/pnas.1712966114
9. Berwick, R., & Chomsky, N. (2011). The biolinguistic program: The current state of its development. In A. M. Di Sciullo & C. Boeckx (Eds.), *The biolinguistic enterprise. New perspectives on the evolution and nature of the human language faculty* (pp. 19–41). Oxford University Press.
10. Berwick, R. C., & Chomsky, N. (2016). *Why only us: Language and evolution.* MIT Press.
11. Bickerton, D. (1990). *Language and species.* University of Chicago Press.
12. Boeckx, C. (2014). What can an extended synthesis do for Biolinguistics? In M. Pina & N. Gontier (Eds.), *The evolution of social communication in primates* (Vol. 1). Springer.
13. Boeckx, C. (2021). Language evolution. Pluralism instead of Minimalism. ABRALIN AO VIVO 2023.
14. Boeckx, C. (2021). Reflections on language evolution: From minimalism to pluralism. In M. Dingemanse & N. J. Enfield (Eds.), *Conceptual foundations of language science 6.* Language Science Press.
15. Brown, R. (1973). *A first language: The early stages.* George Allen & Unwin Ltd.
16. Bybee, J. (2003). Cognitive processes in grammaticalization. In M. Tomasello (Ed.), *The new psychology of language* (Vol. 2). Lawrence Erlbaum.
17. Bybee, J., & Hopper, P. (2001 [2003]). Introduction to frequency and the emergence of linguistic structure. In F. J. Newmeyer (Ed.), *Grammar is grammar and usage is usage.* Project MUSE, John Hopkins University.

J. Dornbierer-Stuart, *The Origins of Language*,
https://doi.org/10.1007/978-3-031-54938-0

18. Cangelosi, A., & Parisi, D. (2001). Computer simulation: A new scientific approach to the study of language evolution. In A. Cangelosi & D. Parisi (Eds.) *Simulating the evolution of language*. Springer.
19. Chomsky, N. (1957). *Syntactic structures*. Mouton.
20. Chomsky, N. (1965). *Aspects of the theory of syntax*. MIT Press.
21. Chomsky, N. (1981). *Lectures on government and binding*. Foris.
22. Chomsky, N. (1995). *The minimalist program*. MIT Press.
23. Christiansen, M., & Chater, N. (2008). Language as shaped by the brain. *Behavioral and Brain Sciences, 31*, 489–558. https://doi.org/10.1017/S0140525X08004998
24. Corballis, M. C. (2002). *From hand to mouth*. Princeton University Press.
25. Croft, W. (2000). *Explaining language change*. Longman.
26. Crozier, G. K. D. (2008). Reconsidering cultural selection theory. *The British Journal for the Philosophy of Science, 59*(3), 455–479. https://doi.org/10.1093/bjps/axn018
27. Crystal, D. (1987). *The Cambridge encyclopedia of language*. Guild Publishing.
28. Cuskley, C. (2020). *Language evolution: A brief overview*. PsyArXiv. https://doi.org/10.31234/osf.io/3y98j
29. Darwin, C. (1859). *On the origin of species*. John Murray.
30. Darwin, C. (1871). *The descent of man*. John Murray.
31. Dawkins, R. (1976). *The selfish gene*. Oxford University Press.
32. Dawkins, R., & Krebs, J. R. (1978). Animal signals: Information or manipulation? In J. R. Krebs & N. B. Davies (Eds.), *Behavioural ecology* (pp. 282–309). Blackwell Scientific Publications.
33. Deacon, T. (1997). *The symbolic species: The co-evolution of language and the brain*. Norton.
34. Deacon, T. (2010). A role for relaxed selection in the evolution of the language capacity. *Proceedings of the National Academy of Sciences, 107*(2), 9000–9006. https://doi.org/10.1073/pnas.0914624107
35. de Boer, B. (2000). Self organization in vowel systems. *Journal of Phonetics, 28*(4), 441–465. https://doi.org/10.1006/jpho.2000.0125
36. Dediu, D., et al. (2013). Cultural evolution of language. In P. J. Richerson & M. H. Christiansen (Eds.), *Cultural evolution: Society, technology, language and religion* (Vol. 12, Ch. 16, pp. 303–332). MIT Press.
37. de Menocal, P. (1995). Plio-Pleistocene African climate. *Science, 270*, 53–59. https://doi.org/10.1126/science.270.5233.53
38. de Saussure, F. (1916). *Cours de linguistique générale*.
39. Deutscher, G. (2005). *The unfolding of language*. Metropolitan Books.
40. di Pellegrino, G., Fadiga, L., Fogassi, L., Gallese, V., & Rizzolatti, G. (1992). Understanding motor events: A neurophysiological study. *Experimental Brain Research, 91*, 176–180. https://doi.org/10.1007/BF00230027
41. Dunbar, R. (1996). *Grooming, gossip and the evolution of language*. Faber and Faber.
42. Eckert, P. (2003). *The meaning of style*. Stanford University.
43. Evans, N., & Levinson, S. (2009). The myth of language universals: Language diversity and its importance for cognitive science. *Behavioral and Brain Sciences, 32*(5), 429–448. https://doi.org/10.1017/S0140525X0999094X
44. Falk, D. (2004). Prelinguistic evolution in early hominins: Whence motherese? *Behavioral and Brain Sciences, 27*(4), 491–503. https://doi.org/10.1017/s0140525x04000111
45. Federmeier, K. D., & Kutas, M. (1999). A rose by any other name: Long-term memory structure and sentence processing. *Journal of Memory and Language, 41*, 469–495. https://doi.org/10.1006/JMLA.1999.2660
46. Field, J. (2004). *Psycholinguistics*. Routledge.
47. Fisher, S. E. (2017). Evolution of language. Lessons from the genome. *Psychonomic Bulletin & Review, 24*(1), 34–40. https://doi.org/10.3758/s13423-016-1112-8
48. Fitch, W. T. (2013). Musical protolanguage: Darwin's theory of language evolution revisited. In J. Bolhuis & M. Everaert (Eds.), *Birdsong, speech, and language: Exploring the evolution of mind and brain* (Ch. 24, pp. 489–503). MIT Press.

49. Fitch, W. T. (2017). Empirical approaches to the study of language evolution. *Psychonomic Bulletin & Review, 24*, 3–33. https://doi.org/10.3758/s13423-017-1236-5

50. Fitch, W. T. (2018). The biology and evolution of speech: A comparative analysis. *Annual Review of Linguistics, 4*(1), 255–279. https://doi.org/10.1146/annurev-linguistics-011817-045748

51. Fitch, W. T. (2019). Animal cognition and the evolution of human language: Why we cannot focus solely on communication. *Philosophical Transactions of the Royal Society B, 375*(1789), 20190046. https://doi.org/10.1098/rstb.2019.0046

52. Freeberg, T. M., Dunbar, R. I. M., & Ord, T. J. (2012). Social complexity as a proximate and ultimate factor in communicative complexity. *Philosophical Transactions of the Royal Society B, 367*, 1785–1801. https://doi.org/10.1098/rstb.2011.0213

53. Friederici, A. D. (2002). Towards a neural basis of auditory sentence processing. *Trends in Cognitive Sciences, 6*(2), 78–84. https://doi.org/10.1016/S1364-6613(00)01839-8

54. Friederici, A. D. (2011). *Neuropsychologische Grundlagen der Sprachentwicklung*. Max-Planck-Institut für Kognitions- und Neurowissenschaften.

55. Friederici, A. D. (2017). *Language in our Brain*. MIT Press.

56. Friederici, A. D., Friedrich, M., & Weber, C. (2002). Neural manifestation of cognitive and precognitive mismatch detection in early infancy. *Neuroreport, 13*(10), 1251–1254. https://doi.org/10.1097/00001756-200207190-00006

57. Friederici, A. D., Friedrich, M., & Christophe, A. (2007). Brain responses in 4-month-old infants are already language specific. *Current Biology, 17*(14), 1208–1211. https://doi.org/10.1016/j.cub.2007.06.011

58. Friedrich, M., Weber, C., & Friederici, A. D. (2004). Electro-physiological evidence for delayed mismatch response in infants at-risk for specific language impairment. *Psychophysiology, 41*(5), 772–782. https://doi.org/10.1111/j.1469-8986.2004.00202.x

59. Fromkin, V., & Rodman, R. (1983). *An introduction to language* (3rd ed.). CBS College Publishing.

60. Fujita, H., & Fujita, K. (2021). Human language evolution: A view from theoretical linguistics on how syntax and the lexicon first came into being. *Primates, 63*(5), 403–415. https://doi.org/10.1007/s10329-021-00891-0

61. Ghazanfar, A., & Rendall, D. (2008). Evolution of human vocal production. *Current Biology, 18*(11), R457. https://doi.org/10.1016/j.cub.2008.03.030

62. Goodall, J. (1986). *The chimpanzees of Gombe: Patterns of behaviour*. Cambridge.

63. Gruber, T., Muller, M. N., Strimling, P., Wrangham, R., & Zuberbühler, K. (2009). Wild chimpanzees rely on cultural knowledge to solve an experimental honey acquisition task. *Current Biology, 19*(21), 1806–1810. https://doi.org/10.1016/j.cub.2009.08.060

64. Gull, T. (2021). Mensch und Schimpans. *UZH Magazin, 2*(21), 32–37.

65. Gull, T., & Nickl, R. (2021). Anthropologie: Modernes Hirn. *UZH Magazin, 2*(21), 28.

66. Günther, F., Dudschig, C., & Kaup, B. (2017). Symbol grounding without direct experience. *Cognitive Science, 42*(2), 336–374. https://doi.org/10.1111/cogs.12549

67. Hall, J. K., Cheng, A., & Carlson, M. T. (2006). Reconceptualizing multicompetence as a theory of language knowledge. *Applied Linguistics, 27*(2), 220–240. https://doi.org/10.1093/APPLIN/AML013

68. Harari, Y. N. (2014). *Sapiens: A brief history of humankind*. Harvill Secker.

69. Hartley, A. F. (1982). *Linguistics for language learners*. The Macmillan Press.

70. Hauser, M. D., Chomsky, N., & Fitch, W. T. (2002). The faculty of language: What is it, who has it, and how did it evolve? *Science, 298*, 1569–1579. https://doi.org/10.1126/science.298.5598.1569

71. Hayashi, M. (2015). Perspectives on object manipulation and action grammar for percursive actions in primates. *Philosophical Transactions of the Royal Society B, 370*(1682), 20140350. https://doi.org/10.1098/rstb.2014.0350

72. Heine, B., & Kuteva, T. (2007). *The genesis of grammar*. Oxford University Press.

73. Herder, J. G. (1891/1966). On the origin of language. In J. H. Moran & A. Gode (Eds.), *On the origin of language*. University of Chicago Press.

74. Hirsh-Pasek, K., & Golinkoff, R. (1996). *The origins of grammar: Evidence from early language comprehension.* MIT Press.
75. Hockett, C. (1960). The origin of speech. *Scientific American, 203*, 88–111. https://doi.org/10.1038/scientificamerican0960-88
76. Hooper, J. B. (1976). Word frequency in lexical diffusion and the source of morphophonological change. In W. Christie (Ed.), *Current progress in historical linguistics* (pp. 96–105). Amsterdam.
77. Ibbotson, P., & Tomasello, R. (2016). Evidence Rebuts Chomsky's theory of language learning. *Scientific American* 2023, September 7.
78. Imai, M., & Kita, S. (2014). The sound symbolism bootstrapping hypothesis for language acquisition and language evolution. *Philosophical transactions of the Royal Society B, 369*, 20130298. https://doi.org/10.1098/rstb.2013.0298
79. Inoue, S., & Matsuzawa, T. (2007). Working memory of numerals in chimpanzees. *Current Biology, 17*(23), 1004–1005. https://doi.org/10.1016/j.cub.2007.10.027
80. Jackendoff, R. (1999). Some possible stages in the evolution of the language capacity. *Trends in Cognitive Sciences 3*, 272–79.
81. Jackendoff, R. (2002). *Foundations of language: Brain, meaning, grammar, evolution.* Oxford University Press.
82. Jackendoff, R. (2003). Précis of Foundations of language: Brain, meaning, grammar, evolution. *Behavioral and Brain Sciences, 26*, 651–707. https://doi.org/10.1017/S0140525X03000153
83. Jespersen, O. (1922). Language, its nature, development, and origin. *The American Journal of Philology, 43*(4), 370–373.
84. Kamhi, A. G. (1986). The elusive first word: The importance of the naming insight for the development of referential speech. *Journal of Child Language, 13*(1), 155–161. https://doi.org/10.1017/S0305000900000362
85. Kendon, A. (2016). Reflections on the "gesture-first" hypothesis of language origins. *Psychonomic Bulletin & Review, 24*, 163–170. https://doi.org/10.3758/s13423-016-1117-3
86. Kirby, S., & Hurford, J. R. (2002). The emergence of linguistic structure: An overview of the iterated learning model. In A. Cangelosi & D. Parisi (Eds.), *Simulating the evolution of language* (pp. 121–147). Springer.
87. Kirby, S., Dowman, M., & Griffiths, T. L. (2007). Innateness and culture in the evolution of language. *Proceedings of the National Academy of Sciences, 104*(12), 5241–5245. https://doi.org/10.1073/pnas.0608222104
88. Kuhl, P. K., & Miller, J. D. (1975). Speech perception by the chinchilla: Voiced-voiceless distinction in alveolar plosive consonants. *Science, 190*(4209), 69–72. https://doi.org/10.1126/science.1166301
89. Lemasson, A., Gautier, J.-P., & Hausberger, M. (2004). Vocal similarities and social bonds Campbell's monkey (Cercopithecus campbelli). *Comptes Rendus Biologies, 326*(12), 1185–1193. https://doi.org/10.1016/j.crvi.2003.10.005
90. Lenneberg, E. H. (1967). *Biological foundations of language.* Wiley.
91. Leroux, M., & Townsend, S. W. (2020). Call combinations in great apes and the evolution of syntax. *Animal Behavior and Cognition, 7*(2), 131–139. https://doi.org/10.26451/abc.07.02.07.2020
92. Levelt, W. (1989). *Speaking: From intention to articulation.* The MIT Press.
93. Levelt, W. (1993). The architecture of normal spoken language use. In G. Blanken, J. Dittman, H. Grimm, J. C. Marshall, & C.-W. Wallesch (Eds.), *Linguistic disorders and pathologies: An international handbook* (pp. 1–15). Walter de Gruyter.
94. Levelt, W. (1995). The ability to speak: From intentions to spoken words. *European Review, 3*(1), 13–23. https://doi.org/10.1017/S1062798700001290
95. Levinson, S. C. (1983). *Pragmatics.* Cambridge University Press.
96. Liberman, A. M., Cooper, F. S., Shankweiler, D. P., & Studdert-Kennedy, M. (1967). Perception of the speech code. *Psychological Review, 74*(6), 431–461. https://doi.org/10.1037/h0020279

97. Lichtheim, L. (1885). Über Aphasie. *Deutsches Archiv für klinische Medizin, 36*, 204–268.
98. Lieberman, P. (1984). *The biology and evolution of language.* Harvard University Press.
99. Lieberman, P. (1992). On the evolution of human language. In J. Hawkins, & M. Gell-Mann (Eds.) *The evolution of human languages.* Avalon.
100. Lust, B. (2006). *Child language: Acquisition and growth.* Cambridge University Press.
101. Mampe, B., Friederici, A., Christophe, A., & Wermke, K. (2009). Newborns' cry melody is shaped by their native language. *Current Biology, 19*(23), 1994–1997. https://doi.org/10.1016/j.cub.2009.09.064
102. Marler, P., & Sherman, V. (1985). Innate differences in singing behaviour of sparrows reared in isolation. *Animal Behaviour, 33*(1), 57–71. https://doi.org/10.1016/S0003-3472(85)80120-2
103. Maslin, M. A., Brierley C. M., Milner, A. M., Schultz, S., Trauth, M. H., & Wilson, K. E. (2014). East African climate pulses and early human evolution. *Quaternary Science Reviews, 101*, 1–17. https://doi.org/10.1016/j.quascirev.2014.06.012
104. Matsuzawa, T. (1996). Chimpanzee intelligence in nature and in captivity: Isomorphism of symbol use and tool use. In J. Goodall, J. Itani, & W. Foundation (Authors) & W. McGrew, L. Marchant, & T. Nishida (Eds.), *Great ape societies* (pp. 196–210). Cambridge University Press.
105. Mehler, J., Jusczyk, P., Lambertz, G., Halsted, N., Bertoncini, J., & Amiel-Tison, C. (1988). A precursor of language acquisition in young infants. *Cognition, 29*(2), 143–178. https://doi.org/10.1016/0010-0277(88)90035-2
106. Mendívil-Giró, J.-L. (2018). Is Universal Grammar ready for retirement? A short review of a longstanding misinterpretation. *Journal of Linguistics, 54*(2018), 859–888. https://doi.org/10.1017/S0022226718000166
107. Mendívil-Giró, J.-L. (2019). Did language evolve through language change? On language change, language evolution and grammaticalization theory. *Glossa: A Journal of General Linguistics, 4*(1), 124. 1–30. https://doi.org/10.5334/gigl.895
108. Meyerhoff, M. (2018). *Introducing sociolinguistics* (3rd ed.). Routledge.
109. Mitchell, R., & Myles, F. (2004). *Second language learning theories.* Arnold.
110. Miyagawa, S. (2017). Integration hypothesis: A parallel model of language development in evolution. In S. Watanabe, M. Hofman, & T. Shimizu (Eds.), *Evolution of the brain, cognition, and emotion in vertebrates.* Brain Science. Springer.
111. Newmeyer, F. J. (2003). *Grammar is grammar and usage is usage.* John Hopkins University.
112. Nickl, R. (2021). Actionfilme für Affen. *UZH Magazin, 2*(21), 38.
113. Nölle, J. (2014). A co-evolved continuum of language, culture and cognition: Prospects of interdisciplinary research. *Studies about Languages,* 2014 No. 25. https://doi.org/10.5755/j01.sal.0.25.8504
114. Nölle, J., Hartmann, S., & Tinits, P. (2020). Language evolution research in the year 2020. *Language Dynamics and Change, 10*, 3–26. https://doi.org/10.1163/22105832-bja10005
115. Odling-Smee, J., Laland, K. N., & Feldman, M. W. (2003). Niche construction: The neglected process in evolution. *Monographs in Population Biology, 37*. https://doi.org/10.1515/9781400847266
116. Osvath, M., & Osvath, H. (2008). Chimpanzee (Pan troglodytes) and orangutan (Pongo abelii) forethought: Self-control and pre-experience in the face of future tool use. *Animal Cognition, 11*, 661–674. https://doi.org/10.1007/s10071-008-0157-0
117. Ouattara, K., Lemasson, A., & Zuberbühler, K. (2009a). Campbell's monkeys use affixation to alter call meaning. *PLoS ONE, 4*(11), e7808. https://doi.org/10.1371/journal.pone.0007808
118. Ouattara, K., Lemasson, A., & Zuberbühler, K. (2009b). Campbell's monkeys concatenate vocalizations into context-specific call sequences. *Proceedings of the National Academy of Sciences, 106*(51), 22026–22031. https://doi.org/10.1073/pnas.0908118106
119. Oudeyer, P., & Kaplan, F. (2007). Language evolution as a Darwinian process: Computational studies. *Cognitive Processing, 8*, 21–35. https://doi.org/10.1007/s10339-006-0158-3
120. Pannekamp, A., Weber, C., & Friederici, A. D. (2006). Prosodic processing at the sentence level in infants. *NeuroReport, 17*(6), 675–678. https://doi.org/10.1097/00001756-200604240-00024

121. Pepperberg, I. M. (1999). *The Alex studies: Cognitive and communicative abilities of grey parrots.* Harvard University Press. https://doi.org/10.2307/j.ctvk12qc1

122. Perniss, P., & Vigliocco, G. (2014). The bridge of iconicity: From a world of experience to the experience of language. *Philosophical Transactions of the Royal Society B Biological Sciences, 369,* 20130300. https://doi.org/10.1098/rstb.2013.0300

123. Pinker, S. (1994). *The language instinct.* Penguin Books.

124. Pinker, S., & Jackendoff, R. (2004). The faculty of language: What's special about it? *Cognition, 95*(2), 201–236. https://doi.org/10.1016/j.cognition.2004.08.004

125. Poliva, O. (2015). From where to what: A neuroanatomically based evolutionary model of the emergence of speech in humans. *F1000Research* 4:67. https://doi.org/10.12688/f1000research.6175.1

126. Poliva, O. (2016). From mimicry to language: A neuroanatomically based evolutionary model of the emergence of vocal language. *Frontiers in Neuroscience, 10,* Art 307. https://doi.org/10.3389/fnins.2016.00307

127. Poliva, O. (2017). From where to what: A neuroanatomically based evolutionary model of the emergence of speech in humans. *F1000Research* 6:67. https://doi.org/10.12688/f1000research.6175.3

128. Potts, R. (1998). Environmental hypotheses of hominin evolution. *Yearbook of Physical Anthropology, 41,* 93–136. https://doi.org/10.1002/(sici)1096-8644(1998)107:27+%3c93::aid-ajpa5%3e3.0.co;2-x

129. Progovac, L. (2016). A gradualist scenario for language evolution: Precise linguistic reconstruction of early human (and Neandertal) grammars. *Frontiers in Psychology, 7,* Article 1714. https://doi.org/10.3389/fpsyg.2016.01714

130. Radford, A. (1990). *Syntactic theory and the acquisition of English syntax.* Basil Blackwell.

131. Ramachandran, V. S., & Hubbard, E. M. (2001). Synaesthesia—A window into perception, thought and language. *Journal of Consciousness Studies, 8*(12), 3–34.

132. Richerson, P. J., Boyd, R., & Henrich, J. (2010). Gene-culture coevolution in the age of genomics. *Proceedings of the National Academy of Sciences, 107*(2), 8985–8992. https://doi.org/10.1073/pnas.0914631107

133. Rizzolatti, G., & Arbib, M. A. (1998). Language within our grasp. *Trends in Neurosciences, 21*(5), 188–194. https://doi.org/10.1016/s0166-2236(98)01260-0

134. Roberts, S. G. (2018). Robust, causal, and incremental approaches to investigating linguistic adaptation. *Frontiers in Psychology, 9,* 327602. https://doi.org/10.3389/fpsyg.2018.00166

135. Roberts, S. G., et al. (2020). CHIELD: The causal hypotheses in evolutionary database. *Journal of Language Evolution, 5*(1), 1–20. https://doi.org/10.1093/jole/lzaa001

136. Rumelhart, D., & McClelland, J. (1986). On learning the past tense of English verbs. In J. McClelland & D. Rumelhart (Eds.), *Parallel distributed processing: Explorations in the microstructure of cognition* (Vol. 2: Psychological and biological models, pp. 216–271). MIT Press.

137. Schmid, H.-J. (2012). Linguistic theories, approaches and methods. In M. Middeke, T. Müller, C. Wald, & H. Zapf (Eds.), *English and American Studies.* J. B. Metzler.

138. Scott-Phillips, T. C. (2006). Why talk? Speaking as selfish behaviour. In A. Cangelosi, et al. (Eds.). *The Evolution of Language: Proceedings of the 6th International Conference on the Evolution of Language* (pp. 299–306). Singapore: World Scientific.

139. Scott-Phillips, T. C. (2014). *Speaking our minds.* Palgrave Macmillan.

140. Seyfarth, R. M., Cheney, D. L., & Marler, P. (1980). Monkey responses to three different alarm calls: Evidence of predator classification and semantic communication. *Science, 210,* 801–803. https://doi.org/10.1126/science.7433999

141. Skinner, B. F. (1957). *Verbal behavior.* Appleton-Century-Crofts.

142. Smith, K. (2006). Cultural evolution of language. In E-CÂÂ. K. Brown (Ed.), *Encyclopedia of language & linguistics* (2nd ed., pp. 315–322).

143. Smith, K., Brighton, H., & Kirby, S. (2003). Complex systems in language evolution: The cultural emergence of compositional structure. *Advances in Complex Systems, 6*(4), 537–558. https://doi.org/10.1142/S0219525903001055

144. Stein, J. F. (2003). Why did language develop? *International Journal of Pediatric Otorhinolaryngology, 67*(Supplement 1), 131–135. https://doi.org/10.1016/j.ijporl.2003.08.011
145. Tallerman, M. (2007). Did our ancestors speak a holistic protolanguage? *Lingua, 117*, 579–604. https://doi.org/10.1016/j.lingua.2005.05.004
146. Tomasello, M. (2003). *Constructing a language: A usage-based theory of language acquisition.* Harvard University Press.
147. Tremblay, P., & Dick, A. S. (2016). November). Broca and Wernicke are dead, or moving past the classic model of neurobiology. *Brain and Language, 162*, 60–71. https://doi.org/10.1016/j.bandl.2016.08.004
148. Wacewicz, S., & Zywiczynski, P. (2015). Language evolution: Why Hockett's design features are a non-starter. *Biosemiotics, 8*, 29–46. https://doi.org/10.1007/s12304-014-9203-2
149. Wei, X., Adamson, H., Schwendemann, M., Goucha, T., Friederici, A. D., & Anwander, A. (2023). *Native language differences in the structural connectome of the human brain.* Max Planck Institute for Human Cognitive and Brain Science, Department of Neuropsychology.
150. Wernicke, C. (1874). *Der aphasische Symptomencomplex: eine psychologische Studie auf anatomischer Basis.* Cohn & Weigert.
151. Wray, A. (1998). Protolanguage as a holistic system for social interaction. *Language and Communication, 18*(1), 47–67. https://doi.org/10.1016/S0271-5309(97)00033-5
152. Wray, A. (2000). Holistic utterances in protolanguage: The link from primates to humans. In C. Knight, M. Studdert-Kennedy, & J. R. Hurford (Eds.), *The evolutionary emergence of language: Social function and the origin of linguistic form (CUP)* (pp. 285–302). Cambridge University Press.
153. Yamamoto, S., Humle, T., & Tanaka, M. (2012). Chimpanzees' flexible targeted helping based on an understanding of conspecifics' goals. *Proceedings of the National Academy of Sciences, 109*(9), 3588–3592. https://doi.org/10.1073/pnas.1108517109
154. Yuan, S., & Fisher, C. (2009). "Really? She blicked the baby?" Two-year-olds learn combinatorial facts about verbs by listening. *Psychological Science, 20*(5), 619–626. https://doi.org/10.1111/j.1467-9280.2009.02341.x
155. Zuidema, W. (2005). *The major transitions in the evolution of language.* University of Edinburgh.
156. Zuidema, W., & de Boer, B. (2018). The evolution of combinatorial structure in language. *Current Opinion in Behavioural Sciences, 21*, 138–144. https://doi.org/10.1016/j.cobeha.2018.04.011

Websites

157. ABRALIN AO VIVO. (2023). Cedric Boeckx. Language Evolution. Pluralism instead of Minimalism. 19.06.2021. https://aovivo.abralin.org/en/lives/cedric-boeckx/
158. Scientific American. (2023). Evidence Rebuts Chomsky's Theory of Language Learning. September 7, 2016. By Paul Ibbotson, Michael Tomasello. https://www.scientificamerican.com/article/evidence-rebuts-chomsky-s-theory-of-language-learning/
159. The Causal Hypotheses in Evolutionary Linguistics Database (CHIELD). (2023). https://correlation-machine.com/CHIELD/
160. Wikipedia. (2023). Cultural selection theory. https://en.wikipedia.org/wiki/Cultural_selection_theory
161. Wikipedia. (2023). Gesture. https://en.wikipedia.org/wiki/Gesture#Neurology
162. Wikipedia. (2023). Grammaticalisation. https://en.wikipedia.org/wiki/Grammaticalization
163. Wikipedia. (2023). Neanderthal. https://en.wikipedia.org/wiki?curid=27298083#Language

Illustrations

164. "Australopithecus family" by Mauricio Antón.
165. "Muscular attachments of the hyoid bone" by William Duffy.
166. "Position of the hyoid bone" by William Duffy.
167. "Speech areas of the brain" by William Duffy.
168. "The Emergence of Language" by Leo Cullum, CartoonStock Ltd.
169. "Vocal tract of the chimp and modern human" by William Duffy.

Index

© The Editor(s) (if applicable) and The Author(s), under exclusive licence to Springer
Nature Switzerland AG 2024
J. Dornbierer-Stuart, *The Origins of Language*,
https://doi.org/10.1007/978-3-031-54938-0

Milton Keynes UK
Ingram Content Group UK Ltd.
UKHW021024251024
450108UK00001B/23

9 783031 549373